The AP Physics C Companion

Mechanics

Dan Fullerton

Physics Teacher
Irondequoit High School

Adjunct Professor
Microelectronic Engineering
Rochester Institute of Technology

Credits

Thanks To:

Deborah Lynn, Bob Enck, Robyn MacBride, Stephanie Byers, Mike Powlin, Brad Allen, Rob Spencer, Peter Geschke, Cricket Fullerton, Karen Finter, Joe Kunz, Monica Owens, Paul Sedita, Rich Randall, Mike Madden, Mike Cronmiller, Nathan Stack, Emani Clark, and Chris VanKerkhove

Silly Beagle Productions
Webster, NY
Internet: www.sillybeagle.com
E-Mail: info@sillybeagle.com

Cover Design by Rick Menard

Interior Illustrations by Dan Fullerton, Jupiterimages and NASA unless otherwise noted.
All images and illustrations © 2017 Jupiterimages Corporation and Dan Fullerton unless otherwise noted.

Edited by Joe Kunz

Sales and Ordering Information
http://www.aplusphysics.com/apcm
sales@sillybeagle.com
Volume discounts available.
Electronic editions available.

Printed in the United States of America
ISBN: 978-0-9907243-4-6

1 2 3 4 5 6 7 8 9 0 9 8 7 6 5 4 3

Silly Beagle Productions

Welcome to The AP Physics C Companion - Mechanics. This book is your essential physics companion to complement AP Physics C lectures and physics textbooks, as well as traditional University Physics courses focusing on mechanics (typically University Physics I). It is designed to be used throughout the course as a supplement to traditional materials, endeavoring to highlight fundamental principles and applications, introduce clarifying examples and sample problems, and serve as a study guide for AP exam preparation.

In addition to this book, you'll find tons more resources available to assist you in your studies on the APlusPhysics.com website. There you'll have access to video mini-lessons explaining fundamental concepts for the course, detailed study guides condensing the key information from each unit, a question and answer discussion board to obtain help when you get stuck, and most importantly, a meeting place where you can interact with other students around the world in the same situation as you.

Physics isn't easy, but with hard work, dedication, and the right resources, you can make your own success and open doors to a tremendous variety of rewarding opportunities. I wish you all the enjoyment and success in the world. Make each day a great day!

Though many fantastic teachers and students have contributed to this work, any errors you find are mine and mine alone. Please let me know should you find any, and I'll do my best to keep the latest version of this book as error-free as possible. Thank you for your support!

Dedication

To those who look before them and choose their own path.
To those who draw their own conclusions.
To those for whom loyalty is a way of life.
To the protectors and the defenders.
Honor Super Omnia.

To the Cannons, Fullertons, Kilburns and Langs, who taught me first to dream big,
then helped me believe I could achieve those dreams.

Make Each Day A Great One!

Table of Contents

Chapter 1: Introduction

© HaywireMedia | stock.adobe.com

Objectives

1. Recognize the questions of physics.

2. List several disciplines within the study of physics.

3. Describe the key topics covered in AP Physics C.

4. Define matter, mass, work and energy.

5. Recognize the intent and depth of this book as a companion resource to be used in conjunction with active learning practices such as hands-on exploration, discussion, debate, and deeper problem solving.

Physics is many things to many different people. If you look up physics in the dictionary, you'll probably learn physics has to do with matter, energy, and their interactions. But what does that really mean? What is matter? What is energy? How do they interact? And most importantly, why do we care?

Physics, in many ways, is the answer to the favorite question of most 2-year-olds: "Why?" What comes after the why really doesn't matter. If it's a "why" question, chances are it's answered by physics: Why is the sky blue? Why does the wind blow? Why do we fall down? Why does my teacher smell funny? Why do airplanes fly? Why do the stars shine? Why do I have to eat my vegetables? The answers to all these questions, and many more, ultimately reside in the realm of physics.

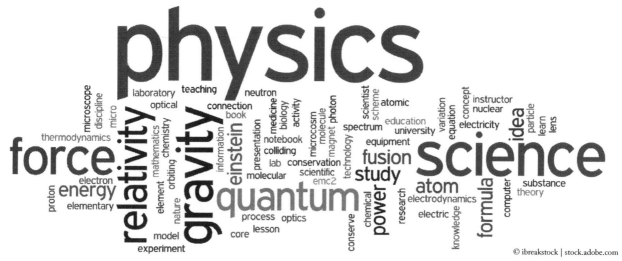

Matter, Systems, and Mass

If physics is the study of matter, then we probably ought to define matter. **Matter,** in scientific terms, is anything that has mass and takes up space. In short, then, matter is anything you can touch – from objects smaller than electrons to stars hundreds of times larger than the sun. From this perspective, physics is the mother of all science. From astronomy to zoology, all other branches of science are subsets of physics, or specializations inside the larger discipline of physics.

In physics, you'll oftentimes talk about **systems** as collections of smaller constituent substructures such as atoms and molecules. The properties and interactions of these substructures determine the properties of the system. In many cases, when the properties of the constituent substructures are not important in understanding the behavior of the system as a whole, the system is referred to as an **object**. For example, a baseball is comprised of many atoms and molecules which determine its properties. For the purpose of analyzing the path of a thrown ball in the air, however, the makeup of its constituent substructures is not important; therefore, we treat the ball as a single object.

So then, what is mass? **Mass** is, in simple terms, the amount of "stuff" an object is made up of. But of course, there's more to the story. Mass is split into two types: inertial mass and gravitational mass. **Inertial mass** is an object's resistance to being accelerated by a force. More massive objects accelerate less than smaller objects given an identical force. **Gravitational mass**, on the other hand, relates to the amount of gravitational force experienced by an object. Objects with larger gravitational mass experience a larger gravitational force.

Confusing? Don't worry! As it turns out, in all practicality, inertial mass and gravitational mass have always been equal for any object measured, even if it's not immediately obvious why this is the case (although with an advanced study of Einstein's Theory of General Relativity you can predict this outcome).

Energy

If it's not matter, what's left? Why, energy, of course. As energy is such an everyday term that encompasses so many areas, an accurate definition can be quite elusive. Physics texts oftentimes define **energy** as the ability or capacity to do work. It's a nice, succinct definition, but leads to another question – what is work? **Work** can also be defined many ways, but a general definition starts with the process of moving an object. If you put these two definitions together, you can vaguely define energy as the ability or capacity to move an object.

Mass-Energy Equivalence

So far, the definition of physics boils down to the study of matter, energy, and their interactions. Around the turn of the 20th century, however, several physicists began proposing a strong relationship between matter and energy. Albert Einstein, in 1905, formalized this with his famous formula $E=mc^2$, which states that the mass of an object, a key characteristic of matter, is really a measure of its energy. This discovery has paved the way for tremendous innovation ranging from nuclear power plants to atomic weapons to particle colliders performing research on the origins of the universe. Ultimately, if traced back to its origin, the source of all energy on earth is the conversion of mass to energy!

Scope of Mechanics

The study of **mechanics**, specifically, deals with the motion of objects, the forces that change the motion of objects, and how these objects interact with other objects and their surrounding environments. In modern times, mechanics has been broken down further into two classifications: **classical mechanics**, which is based on Sir Isaac Newton's laws and formulations, and **quantum mechanics**, which is based on exploration of the wave properties of subatomic particles. As a companion to introductory physics courses, this book will focus solely on classical mechanics.

We'll begin with a review of the math required throughout the course, then dive into a study of **kinematics**, the study of moving objects. We'll then progress to **dynamics**, in which we explore how forces affect the motion of objects. Next, we'll analyze ways in which forces change the energy of objects as they do work on them, and how that energy can translate to various forms. From there, we can look at how moving objects interact with each other through collisions and explosions. Then we'll expand our look at motion by exploring not just motion through space, but motion of rotating and revolving objects. This will provide a great background for exploring repetitive, or oscillating, motion. Finally, we'll end with an exploration of gravity as we combine kinematics, dynamics, momentum, rotation, and oscillations to understand not only how objects attract each other through gravitation, but also how gravity affects the motion of planetary bodies.

It's important to keep in mind, however, that although we're covering a ton of information, there are entire worlds of exploration that we just don't have time to touch on in a course like this. Other areas under the general umbrella of mechanics that aren't covered in this book include acoustics, fluids, soil mechanics, biomechanics, relativistic mechanics, and many others. Further, this is an introduction to these topics. If you enjoy what you study in your mechanics course, be aware that there are ample opportunities to explore any (or all) of these topics in infinitely more depth!

How to Use This Book

This book is not intended as a replacement for a traditional textbook, nor has it been written as a replacement for the popular review books intended to help students improve their AP Exam scores. Instead, this book is intended as a companion book, useful throughout the course in conjunction with attending class, reading the standard textbook, engaging in hands-on laboratory activities, and actively participating in individual and small group problem solving.

So why do you need this book with all these other learning methods available? For starters, traditional textbooks can be challenging to read and understand. I find that my textbooks make much more sense AFTER I have learned the material they contain. I also find, as a teacher, that many of my students don't read the textbook. This isn't always the fault of the textbook, as a focus on absolute accuracy in wording and presentation to set the stage for any future studies in physics many times leads to very long and complicated explanations which are challenging for beginning students to follow for any extended length of time.

In traditional learning environments, this gap is meant to be cleaned up in the lecture halls, but lectures from even the best of professors often go by way too quickly, and often there isn't adequate time to go through detailed application of principles to problem solving.

The AP Physics C Companion: Mechanics fills that gap. Fundamental concepts are presented as quickly and efficiently as is reasonably possible, and are followed up with hundreds of example problems in which the concepts are applied, with full solutions and problem solving steps laid out in detail with mathematical justifications to assist readers in truly understanding the material. Common and popular problems and derivations are presented as simply as possible, and numerous diagrams are included to illustrate concepts, highlight real-world applications, and keep the book fun and light.

Further, The AP Physics C Companion: Mechanics is designed to tie in directly with the APlusPhysics.com website, where students will find many more resources to assist them in their studies, including detailed study guides for each unit as well as free video mini-lessons covering the fundamental topics of the course.

And with that, allow me to formally welcome you to the world of physics and this book. I truly hope you find it a valuable resource, and you enjoy your studies as much as I enjoyed putting this together for you.

— Dan Fullerton

Chapter 2: Background Skills

"Being kidnapped and abused by the undead was worse than calculus, but not by a wide margin."

—Thomm Quackenbush

Objectives

1. Represent numbers in scientific notation with appropriate significant figures.
2. Make both accurate and precise measurements.
3. Utilize the metric system to understand the size and order of magnitude of measurements.
4. Verify calculations and formulas using dimensional analysis.
5. Represent physical quantities appropriately with vectors and scalars.
6. Manipulate vectors through addition, subtraction, and scalar and vector multiplication.
7. Graphically present and analyze data to determine relationships between variables.
8. Recognize the slope of a function as the derivative of that function.
9. Recognize the area under a function as the integral of that function.
10. Calculate derivatives of polynomial, trigonometric, exponential, and logarithmic functions.
11. Calculate indefinite and definite integrals of polynomial, trigonometric, exponential, and logarithmic functions.

Building a house without knowing how to use a hammer and saw seems like a recipe for failure — unless, that is, you're building the house for those "special" relatives of yours that you're always trying to forget. Having knowledge of and access to the proper tools doesn't guarantee success in your endeavor, but it definitely improves the odds. In similar fashion, success in calculus-based physics is much more likely for those with knowledge of the tools and skills frequently utilized in the course.

Although physics and mathematics are not the same thing, they are in many ways closely related. Just like English is the language of this paragraph, mathematics is the language of physics. A solid understanding of some basic math concepts will allow you to communicate and describe the physical world both efficiently and accurately.

In this chapter you'll review these key skills. Some of these you may be familiar with, while others, such as vectors, scalars, and calculus, may be new. This chapter will focus on providing the practical "need to know" information required for success on the AP Physics C exam, leaving the theoretical derivations and more accurate conceptual explanations in the hands of the experts.

Significant Figures

Significant Figures (or sig figs, for short) represent a manner of showing which digits in a number are known to some level of certainty. But how do you know which digits are significant? There are some rules to help with this. If you start with a number in scientific notation:

- All non-zero digits are significant.
- All digits between non-zero digits are significant.
- Zeroes to the left of significant digits are not significant.
- Zeroes to the right of significant digits are significant.

When you make a measurement in physics, you want to write what you measured using significant figures. To do this, write down as many digits as you are absolutely certain of, then take a shot at one more digit as accurately as you can. These are your significant figures.

2.1 Q: How many significant figures are in the value 43.74 km?

2.1 A: 4 (four non-zero digits)

2.2 Q: How many significant figures are in the value 4302.5 g?

2.2 A: 5 (All non-zero digits are significant and digits between non-zero digits are significant.)

2.3 Q: How many significant figures are in the value 0.0083s?

2.3 A: 2 (All non-zero digits are significant. Zeroes to the left of significant digits are not significant.)

2.4 Q: How many significant figures are in the value 1.200×10^3 kg?

2.4 A: 4 (Zeroes to the right of significant digits are significant.)

As the focus of this book is building a solid understanding of basic physics concepts and applications, significant figures will not be emphasized in routine problem solving, but realize that in certain environments they can be of the highest importance. For the purposes of the AP Physics C exams, typically 3-4 significant figures will be adequate.

Scientific Notation

Because measurements of the physical world vary so tremendously in size (imagine trying to describe the distance across the United States in units of hair thicknesses), scientists oftentimes use what is known as scientific notation to represent very large and very small numbers. These very large and very small numbers would become quite cumbersome to write out repeatedly. Imagine writing 4,000,000,000,000 over and over again. Your hand would get tired and your pen would rapidly run out of ink! Instead, it's much easier to write this number as 4×10^{12}. Or on the smaller scale, the thickness of the insulating layer (known as a gate dielectric) in the integrated circuits that power computers and other electronics can be less than 0.000000001 m. It's easy to lose track of how many zeros you have to deal with, so scientists instead would write this number as 1×10^{-9} m. See how much simpler life can be with scientific notation?

Scientific notation follows a few simple rules. Start by showing all the significant figures in the number you're describing, with the decimal point after the first significant digit. Then, show your number being multiplied by 10 to the appropriate power in order to give you the correct value.

It sounds more complicated than it is. Let's say, for instance, you want to show the number 300,000,000 in scientific notation (a very useful number in physics), and let's assume you know this value to three significant digits. You would start by writing the three significant digits, with the decimal point after the first digit, as "3.00". Now, you need to multiply this number by 10 to some power in order to get back to the original value. In this case, you multiply 3.00 by 10^8, for an answer of 3.00×10^8. Interestingly, the power you raise the 10 to is exactly equal to the number of digits you moved the decimal to the left as you converted from standard to scientific notation. Similarly, if you start in scientific notation, to convert to standard notation, all you have to do is remove the 10^8 power by moving the decimal point eight digits to the right. Presto, you're an expert in scientific notation! To make life even easier, put your calculator into scientific mode. Cool, huh?

But what do you do if the number is much smaller than one? Same basic idea. Let's assume you're dealing with the approximate radius of an electron, which is 0.00000000000000282 m. It's easy to see how unwieldy this could become. You can write this in scientific notation by writing out three significant digits, with the decimal point after the first digit, as "2.82." Again, you multiply this number by some power of 10 in order to get back to the original value. Because your value is less than 1, you need to use negative powers of 10. If you raise 10 to the power -15, specifically, you get a final value of 2.82×10^{-15} m. In essence, for every digit you moved the decimal place, you add another power of 10. And if you start with scientific notation, all you do is move the decimal place left one digit for every negative power of 10.

2.5 Q: Express the number 0.000470 in scientific notation.
2.5 A: 4.70×10^{-4}

2.6 Q: Express the number 2,870,000 in scientific notation.
2.6 A: 2.87×10^6

2.7 Q: Expand the number 9.56×10^{-3}.
2.7 A: 0.00956

2.8 Q: Expand the number 1.11×10^7.
2.8 A: 11,100,000

Accuracy and Precision

When making measurements of physical quantities, how close the measurement is to the actual value is known as the accuracy of the measurement. Precision, on the other hand, is the repeatability of a measurement. A common analogy involves an archer shooting arrows at the target. The bullseye of the target represents the actual value of the measurement.

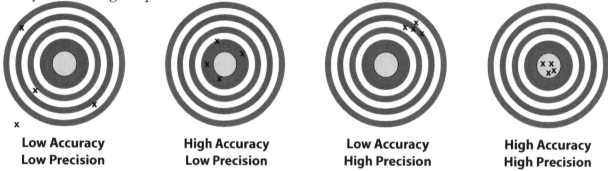

Low Accuracy Low Precision   **High Accuracy Low Precision**   **Low Accuracy High Precision**   **High Accuracy High Precision**

Ideally, measurements in physics should be both accurate and precise.

Metric System

Physics involves the study, prediction, and analysis of real-world phenomena. To communicate data accurately, you must set specific standards for basic measurements. The physics community has standardized what is known as the Système International (SI), which defines seven baseline measurements and their standard units, forming an implementation of the metric system of measurement. The three most common measurement units are meters, kilograms, and seconds, which will be the focus for the majority of this course, which is why this system is sometimes known as the mks system. The fourth SI base unit you'll use in AP Physics C, the ampere, will be introduced in the E&M portion of the course.

The base unit of length in the metric system, the meter, is roughly equivalent to the English yard. For smaller measurements, the meter is divided up into 100 parts, known as centimeters, and each centimeter is made up of 10 millimeters. For larger measurements, the meter is grouped into larger units of 1000 meters, known as a kilometer. The length of a baseball bat is approximately one meter, the radius of a U.S. quarter is approximately a centimeter, and the diameter of the metal in a wire paperclip is roughly one millimeter.

The base unit of mass, the kilogram, is roughly equivalent to two U.S. pounds. A cube of water 10 cm x 10 cm x 10 cm has a mass of 1 kilogram. Kilograms can also be broken up into larger and smaller units, with commonly used measurements of grams (1/1000th of a kilogram) and milligrams (1/1000th of a gram). The mass of a textbook is approximately 2 to 3 kilograms, the mass of a baseball is approximately 145 grams, and the mass of a mosquito is 1 to 2 milligrams. The base unit of time, the second, is likely already familiar. Time can also be broken up into smaller units such as milliseconds (10^{-3} seconds), microseconds (10^{-6} seconds), and nanoseconds (10^{-9} seconds), or grouped into larger units such as minutes (60 seconds), hours (60 minutes), days (24 hours), and years (365.25 days).

The metric system is based on powers of 10, allowing for easy conversion from one unit to another. A chart showing the meaning of commonly used metric prefixes and their notations can be extremely valuable in performing unit conversions.

Prefixes for Powers of 10		
Prefix	Symbol	Notation
tera	T	10^{12}
giga	G	10^{9}
mega	M	10^{6}
kilo	k	10^{3}
deci	d	10^{-1}
centi	c	10^{-2}
milli	m	10^{-3}
micro	μ	10^{-6}
nano	n	10^{-9}
pico	p	10^{-12}

Converting from one unit to another can be easily accomplished if you use the following procedure.

1. Write your initial measurement with units as a fraction over 1.
2. Multiply your initial fraction by a second fraction, with a numerator (top number) having the units you want to convert to, and the denominator (bottom number) having the units of your initial measurement.
3. For any units on the top right-hand side with a prefix, determine the value for that prefix. Write that prefix in the right-hand denominator. If there is no prefix, use 1.
4. For any units on the right-hand denominator with a prefix, write the value for that prefix in the right-hand numerator. If there is no prefix, use 1.
5. Multiply through the problem, taking care to accurately record units. You should be left with a final answer in the desired units.

Let's take a look at a sample unit conversion:

2.9 Q: Convert 23 millimeters (mm) to meters (m).

2.9 A: Step 1. $\dfrac{23mm}{1}$

Step 2. $\dfrac{23mm}{1} \times \dfrac{m}{mm}$

Step 3. $\dfrac{23mm}{1} \times \dfrac{m}{1mm}$

Step 4. $\dfrac{23mm}{1} \times \dfrac{10^{-3}m}{1mm}$

Step 5. $\dfrac{23mm}{1} \times \dfrac{10^{-3}m}{1mm} = 2.3 \times 10^{-2}m$

Now, try some on your own!

2.10 Q: Convert 2.67×10^{-4} m to mm.

2.10 A: $\dfrac{2.67 \times 10^{-4}m}{1} \times \dfrac{1mm}{10^{-3}m} = 0.267mm$

2.11 Q: Convert 14 km/hr to m/s.

2.11 A: $\dfrac{14km}{hr} \times \dfrac{10^{3}m}{1km} \times \dfrac{1hr}{60\min} \times \dfrac{1\min}{60s} = 3.89\,{}^m\!/_s$

2.12 Q: Convert 3,470,000 μs to s.

2.12 A: $\dfrac{3,470,000\mu s}{1} \times \dfrac{10^{-6}s}{1\mu s} = 3.47s$

Analyzing the units involved in a formula or answer can provide considerable insight into whether the formula or answer is correct. This process is known as **dimensional analysis**. Common units may include:

- r (displacement, radius) - meters
- t (time) - seconds
- v (speed, velocity) - meters/second
- a (acceleration) - meters/second²
- F (force) - newton = kilogram•meter/second²
- G (gravitational constant) - Newton•meter²/kilogram²

2.13 Q: Determine which of the following formulas are dimensionally correct.
(A) $v = v_0 t + a$

(B) $d = v_0 t^2 + \frac{1}{2}at^2$

(C) $F = G\dfrac{m_1 m_2}{r^2}$

2.13 A: (C) is the only dimensionally correct formula.

Chapter 2: Background Skills

2.14 Q: A newton per kilogram is equivalent to which of the following units?
(A) s^2
(B) $kg{\cdot}m/s$
(C) m^2/kg
(D) m/s^2
(E) kg/s^2

2.14 A: (D) m/s^2

2.15 Q: Convert the following measurements as shown, maintaining significant figures and showing your final answer in scientific notation.
(A) 25.0 km = _____ m
(B) 8234 mg = _____ kg
(C) $2.46{\times}10^{-7}$ hrs = _____ ms

2.15 A: (A) $2.50{\times}10^{4}$
(B) $8.234{\times}10^{-3}$
(C) $8.856{\times}10^{-1}$

Trigonometry and Triangles

The use of trigonometry, the study of triangles, can be distilled down to the Pythagorean Theorem and definitions of the three basic trigonometric functions. When you know the length of two sides of a right triangle, or the length of one side and a non-right angle, you can solve for all the angles and sides of the triangle. If you can use the definitions of the sine, cosine, and tangent, you'll be fine in physics.

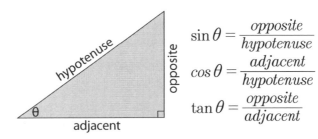

$$\sin\theta = \frac{opposite}{hypotenuse}$$
$$\cos\theta = \frac{adjacent}{hypotenuse}$$
$$\tan\theta = \frac{opposite}{adjacent}$$

If you need to solve for the angles themselves, you can use the inverse trigonometric functions.
$$\theta = \sin^{-1}\left(\frac{opposite}{hypotenuse}\right) = \cos^{-1}\left(\frac{adjacent}{hypotenuse}\right) = \tan^{-1}\left(\frac{opposite}{adjacent}\right)$$

Of course, knowing two sides of a right triangle allows you to find the third side using the Pythagorean Theorem.

$$a^2 + b^2 = c^2$$

Vectors and Scalars

Quantities in physics are used to represent real-world measurements, and therefore physicists use these quantities as tools to better understand the world. In examining these quantities, there are times when just a number, with a unit, can completely describe a situation. These numbers, which have just a magnitude, or size, are known as scalars. Examples of scalars include quantities such as temperature, mass, and time. At other times, a quantity is more descriptive if it also includes a direction. These quantities which have both a magnitude and direction are known as vectors. Vector quantities you may be familiar with include force, velocity, and acceleration.

Most students will be familiar with scalars, but to many, vectors may be a new and confusing concept. By learning just a few rules for dealing with vectors, you'll find that they can be a powerful tool for problem solving.

Vectors are often represented as arrows, with the length of the arrow indicating the magnitude of the quantity, and the direction of the arrow indicating the direction of the vector. Vectors out of the page are shown as dots (as if an arrow were coming toward you), and vectors into the page are shown as X's (as if an arrow is going away from you). In the figure below, vector B has a magnitude greater than that of vector A even though vectors A and B point in the same direction. It's also important to know that vectors can be moved anywhere in space. The positions of A and B could be swapped, and the individual vectors would retain their values of magnitude and direction.

To add vectors A and B below, all you have to do is line them up so that the tip of the first vector touches the tail of the second vector. Then, to find the sum of the vectors, known as the resultant, draw a straight line from the start of the first vector to the end of the last vector. This method works with any number of vectors.

A + B = C

So how do you subtract two vectors? Try subtracting B from A. You can start by rewriting the expression A - B as A + -B. Now it becomes an addition problem. You just have to figure out how to express –B. This is easier than it sounds. To find the opposite of a vector, just point the vector in the opposite direction. Therefore, you can use what we already know about the addition of vectors to find the resultant of A-B.

A + -B = C

You'll learn more about vectors as you go, but before moving on, there are a few basic skills to master. Vectors at angles can be challenging to deal with. By transforming a vector at an angle into two vectors, one parallel to the x-axis and one parallel to the y-axis, you can greatly simplify problem solving. To break a vector up into its components, you can use the basic trig functions.

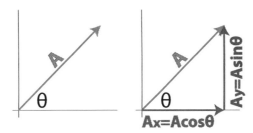

2.16 Q: An airplane flies with a velocity of 750 kilometers per hour, 30° south of east. What is the magnitude of the plane's eastward velocity?
(A) 866 km/h
(B) 650 km/h
(C) 433 km/h
(D) 375 km/h
(E) 0 km/h

2.16 A: (B) The plane's eastward velocity will be the x-component of its velocity vector:

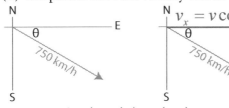

$$v_x = v\cos\theta = \left(750\,\tfrac{km}{hr}\right)\cos\left(30°\right) = 650\,\tfrac{km}{hr}$$

In similar fashion, you can use the components of a vector in order to build the original vector. Graphically, if you line up the component vectors tip-to-tail, the original vector runs from the starting point of the first vector to the ending point of the last vector. To determine the magnitude of the resulting vector algebraically, just apply the Pythagorean Theorem.

2.17 Q: A motorboat, which has a speed of 5.0 meters per second in still water, is headed east as it crosses a river flowing south at 3.3 meters per second. What is the magnitude of the boat's resultant velocity with respect to the starting point?
(A) 1.7 m/s
(B) 3.3 m/s
(C) 5.0 m/s
(D) 6.0 m/s
(E) 8.3 m/s

2.17 A: (D) 6.0 m/s

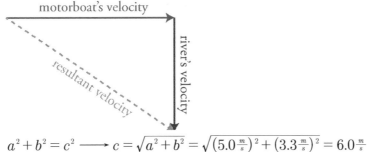

$$a^2 + b^2 = c^2 \longrightarrow c = \sqrt{a^2 + b^2} = \sqrt{\left(5.0\tfrac{m}{s}\right)^2 + \left(3.3\tfrac{m}{s}\right)^2} = 6.0\tfrac{m}{s}$$

Vector Notation

With these tools, you can now begin to work with vectors in three dimensions. Consider a vector **A** in three dimensional space (coordinate axes being x, y, and z). Vector **A** has dimensions of 4 units in the +x direction, 3 units in the +y direction, and 1 unit in the +z direction, as represented below.

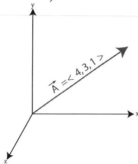

You can recognize this vector **A** in text by noting it is listed in bold. You'll also see vectors noted as having an arrow above them, as shown here: $\vec{A}$. The vector could also be represented analytically by listing its dimensions along the coordinate axes x, y, and z: $\vec{A} = <x, y, z>$. You could write this more specifically as $\vec{A} = <4, 3, 1>$. Becoming more specific, assuming the units are meters, you could write **A** as $\vec{A} = <4m, 3m, 1m>$. You can probably see where the analytic version of writing vector **A** as $\vec{A} = <4m, 3m, 1m>$ can be considerably quicker than drawing a graph every time you want to represent a vector. Formally, this is known as **ordered set notation**, though it is often referred to as **bracket notation**.

Alternately, vectors can be written analytically using **unit vector notation**. Begin by defining a vector of unit length (length equals one) in the x-direction. Let's call this vector $\hat{i}$ (often pronounced "i-hat.") A vector of length 4 meters in the x-direction could be written in bracket notation as $<4m, 0, 0>$, or, using vector notation, this would be $(4m)\hat{i}$. In similar fashion, the unit vector in the y-direction is known as $\hat{j}$, and the unit vector in the z-direction is known as $\hat{k}$. Vector **A**, which we previously represented as $\vec{A} = <4m, 3m, 1m>$, could therefore be written in unit vector notation as $\vec{A} = (4m)\hat{i} + (3m)\hat{j} + (1m)\hat{k}$. From a notation perspective, you can represent the magnitude of a vector by placing absolute value signs around it. The magnitude of vector **A** can be represented as $|\vec{A}|$, or by not using bold text.

The notation looks a bit scary until you get used to it, but the underlying concepts are quite simple, and the analytical notation is quite powerful for vector manipulation. Assume we have vectors **A** and **B** and we want to add them together. We've previously defined vector **A** as $\vec{A} = <4m, 3m, 1m>$, so let's also define vector **B** as $\vec{B} = <-2m, 0, 3m>$. The sum, or resultant, of these two vectors, **R**, can be found by simply adding up the components of the vectors.

$$\vec{R} = \vec{A} + \vec{B}$$
$$\vec{A} = <4m, 3m, 1m>$$
$$+\vec{B} = <-2m, 0m, 3m>$$
$$\overline{\vec{R} = <2m, 3m, 4m>}$$

Alternately, this can be done using unit vector notation:

$$\vec{R} = \vec{A} + \vec{B}$$
$$\vec{A} = (4m)\hat{i} + (3m)\hat{j} + (1m)\hat{k}$$
$$+\vec{B} = (-2m)\hat{i} + (0)\hat{j} + (3m)\hat{k}$$
$$\overline{\vec{R} = (2m)\hat{i} + (3m)\hat{j} + (4m)\hat{k}}$$

2.18 Q: Three "building block" vectors are shown below.

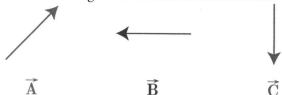

$$\vec{A} \qquad\qquad \vec{B} \qquad\qquad \vec{C}$$

Given the resultant vector $\vec{R}$, determine how to construct the resultant using only the "building block" vectors.

Q1)	Q2)	Q3)
$\vec{R}$ $\vec{R} =$	$\vec{R} =$ $\vec{R}$	$\vec{R}$ $\vec{R} =$
Q4)	Q5)	Q6)
$\vec{R}$ $\vec{R} =$	$\vec{R}$ $\vec{R} =$	$\vec{R}$ $\vec{R} =$

2.18 A:

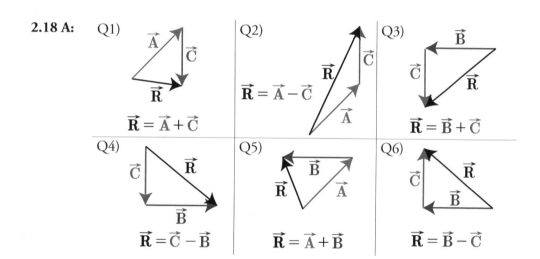

Q1)	Q2)	Q3)
$\vec{A}$ $\vec{C}$ $\vec{R}$ $\vec{R} = \vec{A} + \vec{C}$	$\vec{R}$ $\vec{C}$ $\vec{A}$ $\vec{R} = \vec{A} - \vec{C}$	$\vec{B}$ $\vec{C}$ $\vec{R}$ $\vec{R} = \vec{B} + \vec{C}$
Q4)	Q5)	Q6)
$\vec{C}$ $\vec{R}$ $\vec{B}$ $\vec{R} = \vec{C} - \vec{B}$	$\vec{B}$ $\vec{R}$ $\vec{A}$ $\vec{R} = \vec{A} + \vec{B}$	$\vec{C}$ $\vec{R}$ $\vec{B}$ $\vec{R} = \vec{B} - \vec{C}$

2.19 Q: A frog hops four meters at an angle of 30 degrees north of east. He then hops six meters at an angle of 60 degrees north of west. What is the frog's total displacement from his starting position?

2.19 A: This problem only requires analysis in two dimensions, and can be solved both graphically and analytically.

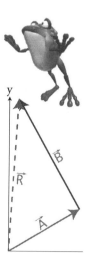

Break up the displacement vectors into their x- and y-components:
$\vec{A} = < 4m\cos 30°, 4m\sin 30° >$
$\vec{B} = < -6m\cos 60°, 6m\sin 60° >$
$\vec{R} = < 4m\cos 30° - 6m\cos 60°, 4m\sin 30° + 6m\sin 60° >$
$\vec{R} = < 0.46m, 7.2m >$

You have solved for the correct answer in terms of x- and y-components, as noted above in bracket notation, but what is the frog's displacement from the starting position in terms of distance and direction?

$$|\vec{R}| = \sqrt{(0.46m)^2 + (7.2m)^2} = 7.21m$$

$$\theta = \tan^{-1}\left(\frac{7.2m}{0.46m}\right) = 86.3° \; North\; of\; East$$

In addition to adding and subtracting vectors, you can take the product of two vectors, analogous to multiplication. There are two types of vector products. The product of two vectors which gives a scalar as its output is known as the **dot product**, or **scalar product**. The product of two vectors which gives a third vector as its output is known as the **cross product**, or **vector product**.

Dot Product

The dot (scalar) product of two vectors uses the component of a given vector in the direction of a second vector. The product of $\vec{A} \cdot \vec{B}$ is the component of vector $\vec{A}$ in the direction of vector $\vec{B}$ times the magnitude of vector $\vec{B}$. Mathematically, you can define the product of $\vec{A} \cdot \vec{B}$ as the magnitude of $\vec{A}$ multiplied by the magnitude of vector $\vec{B}$ multiplied by the cosine of the angle between $\vec{A}$ and $\vec{B}$.

$$\vec{A} \cdot \vec{B} = |\vec{A}||\vec{B}|\cos\theta$$

Calculating the dot product is straightforward if you know the x, y, and z-components of your vectors. You just add up the products of the x-components, the y-components, and the z-components of the original vectors. In unit vector notation:

$$\vec{A} = A_x\hat{i} + A_y\hat{j} + A_z\hat{k}$$
$$\vec{B} = B_x\hat{i} + B_y\hat{j} + B_z\hat{k}$$
$$\vec{A} \cdot \vec{B} = A_xB_x + A_yB_y + A_zB_z$$

Or, in bracket notation:

$$\vec{A} = < A_x, A_y, A_z >$$
$$\vec{B} = < B_x, B_y, B_z >$$
$$\vec{A} \cdot \vec{B} = A_xB_x + A_yB_y + A_zB_z$$

2.20 Q: Find the dot product of the following vectors, and the angle between the vectors.
$$\vec{A} = < 1, 2, 3 >$$
$$\vec{B} = < 3, 2, 1 >$$

2.20 A: The dot product is straightforward: $\vec{A} \cdot \vec{B} = (1 \times 3) + (2 \times 2) + (3 \times 1) = 10$.
You can go back to the definition of the dot product to find the angle between $\vec{A}$ and $\vec{B}$.
$$\vec{A} \cdot \vec{B} = |\vec{A}||\vec{B}|\cos\theta$$
$$\theta = \cos^{-1}\left(\frac{\vec{A} \cdot \vec{B}}{|\vec{A}||\vec{B}|}\right)$$
$$\theta = \cos^{-1}\left(\frac{10}{\sqrt{1^2 + 2^2 + 3^2}\sqrt{3^2 + 2^2 + 1^2}}\right)$$
$$\theta = \cos^{-1}\left(\frac{10}{14}\right)$$
$$\theta = 44.4°$$

Chapter 2: Background Skills

Manipulation of dot products is an important skill in calculus-based physics. The following properties of dot products will be important to know.

- The dot product of perpendicular vectors is always zero.
- The dot product of parallel vectors is the product of their magnitudes.
- Commutative: $\vec{A} \cdot \vec{B} = \vec{B} \cdot \vec{A}$
- Scaling: $k(\vec{A} \cdot \vec{B}) = (k\vec{A}) \cdot \vec{B}$
- Distributive: $(\vec{A} + \vec{B}) \cdot \vec{C} = \vec{A} \cdot \vec{C} + \vec{B} \cdot \vec{C}$
- Product Rule for Derivatives: $\frac{d}{dt}(\vec{A} \cdot \vec{B}) = \frac{d\vec{A}}{dt} \cdot \vec{B} + \frac{d\vec{B}}{dt} \cdot \vec{A}$ (Don't worry if you don't understand this one yet.)

2.21 Q: If $\vec{A} = <-2,3>$ and $\vec{B} = <4, B_y>$, find the value of B_y such that the two vectors are perpendicular.

2.21 A: If the vectors are perpendicular, then their dot product must be zero.
$$\vec{A} \cdot \vec{B} = 0 \xrightarrow[\vec{B}=<4,B_y>]{\vec{A}=<-2,3>} -8 + 3B_y = 0 \longrightarrow B_y = \frac{8}{3}$$

Author's Note: Mathematics is the "language" of physics, much like English is the language of this book. Logical reasoning is written from left to right, in sentence structure, justifying arguments as warranted. In similar fashion, mathematical reasoning in physics is written from left to right, as illustrated in the answer to the previous problem. The arrow to the right shows a jump to the next form of the equation, with any justifications for the jump noted above and below the arrow.

If you were reading the answer to the previous question out loud, you would read it as "A dot B equals zero, which implies that negative eight plus three B subscript y equals zero, which implies that B subscript y equals eight thirds." The first arrow to the right shows the inclusion of justification points above and below the arrow.

For many, this may seem quite strange and awkward as you begin your journey into the realm of physics. For those continuing on into deeper physics and engineering, however, learning to write mathematical reasoning from left to right, and including appropriate justifications, is a great habit to get into, and will be demonstrated liberally in this book.

Cross Product

The cross product, or vector product, of two vectors gives you a vector perpendicular to both whose magnitude is equal to the area of a parallelogram defined by the two initial vectors.
$$\vec{A} \times \vec{B} = \vec{C}$$

Mathematically, you can define the magnitude of the product of $\vec{A} \times \vec{B}$ as the magnitude of $\vec{A}$ multiplied by the magnitude of $\vec{B}$ multiplied by the sine of the angle between $\vec{A}$ and $\vec{B}$.

$$|\vec{A} \times \vec{B}| = |\vec{A}||\vec{B}|\sin\theta$$

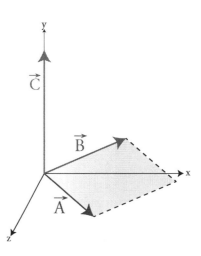

The direction of the cross product vector is given by the right-hand rule. Point the fingers of your right hand in the direction of $\vec{A}$ and then bend your fingers in the direction of $\vec{B}$ (go the short way — try not to break any bones). Your thumb, when extended, will then point in the direction of the cross product vector $\vec{C}$. The cross product is always into or out of the plane formed by the other two vectors.

Calculating the cross product is significantly more involved than dot product calculations. In unit vector notation, the formula for the cross product is:

$$\vec{A} \times \vec{B} = (A_y B_z - A_z B_y)\hat{i} + (A_z B_x - A_x B_z)\hat{j} + (A_x B_y - A_y B_x)\hat{k}$$

That's quite a bit to memorize, but there are some tricks to simplify it. Note that the x-term of the cross product has no dependence on the x terms of the initial vectors... in similar fashion, the y term of the cross product has no dependence on the y terms of the initial vectors, and the z term of the cross product has no dependence on the z terms of the initial vectors.

Alternately, you can recreate the cross product formula by taking a matrix algebra approach. The cross product formula is the determinant of a 3x3 matrix consisting of the unit vectors in the first row, the components of $\vec{A}$ in the second row, and the components of $\vec{B}$ in the third row.

$$\vec{A} \times \vec{B} = \begin{vmatrix} \hat{i} & \hat{j} & \hat{k} \\ A_x & A_y & A_z \\ B_x & B_y & B_z \end{vmatrix}$$

One method of solving this determinant involves the "Rule of Sarrus," which is demonstrated briefly below. Add up the products of the solid diagonals (down and to the right), and subtract the products of the dashed diagonals (down and to the left). For further information on solving a three-by-three matrix, students are encouraged to utilize their Google-Fu and the wonderful resources of the Internet.

$$\vec{A} \times \vec{B} = \begin{array}{ccc|ccc|cc} \hat{i} & \hat{j} & \hat{k} & & & & \\ A_y & A_z & A_x & A_y & A_z & A_x & A_y \\ B_y & B_z & B_x & B_y & B_z & B_x & B_y \end{array}$$

$$\vec{A} \times \vec{B} = (A_y B_z - A_z B_y)\hat{i} + (A_z B_x - A_x B_z)\hat{j} + (A_x B_y - A_y B_x)\hat{k}$$

2.22 Q: Find the cross product of the following vectors:
$\vec{A} = <0, 2, 0>$
$\vec{B} = 2\hat{i}$
$\vec{C} = \vec{A} \times \vec{B} = ?$

2.22 A: First re-write vector **B** in bracket notation for consistency: $\vec{B} = <2, 0, 0>$.
Next, either utilize the formula method or the determinant method to find **C**.
Formula Method
$\vec{C} = \vec{A} \times \vec{B} = (A_y B_z - A_z B_y)\hat{i} + (A_z B_x - A_x B_z)\hat{j} + (A_x B_y - A_y B_x)\hat{k}$
$\vec{C} = (2 \times 0 - 0 \times 0)\hat{i} + (0 \times 2 - 0 \times 0)\hat{j} + (0 \times 0 - 2 \times 2)\hat{k}$
$\vec{C} = -4\hat{k}$

Chapter 2: Background Skills

Determinant Method

$$\vec{C} = \vec{A} \times \vec{B} = \begin{array}{cc} A_y & A_z \\ B_y & B_z \end{array} \begin{vmatrix} \hat{i} & \hat{j} & \hat{k} \\ A_x & A_y & A_z \\ B_x & B_y & B_z \end{vmatrix} \begin{array}{cc} A_x & A_y \\ B_x & B_y \end{array}$$

$$\vec{C} = \vec{A} \times \vec{B} = \begin{array}{cc} 2 & 0 \\ 0 & 0 \end{array} \begin{vmatrix} \hat{i} & \hat{j} & \hat{k} \\ 0 & 2 & 0 \\ 2 & 0 & 0 \end{vmatrix} \begin{array}{cc} 0 & 2 \\ 2 & 0 \end{array}$$

$$\vec{C} = (\hat{k})(0)(0) - (\hat{k})(2)(2) = -4\hat{k}$$

Note that you obtain the same answer regardless of method. This sample is also useful for demonstrating further properties of the cross product. The initial vectors, with unit length two in the x-direction and unit length two in the y-direction, define a parallelogram of area four. The magnitude of the cross product is four. The cross product should also be perpendicular to both our initial vectors. The original vectors were along the x- and y-axes. The cross product is perpendicular to both, along the z-axis. Finally, you can verify your answer by checking the magnitude of the cross product.

$$|\vec{C}| = |\vec{A} \times \vec{B}| = |\vec{A}||\vec{B}|\sin\theta = (2)(2)\sin 90° = 4.$$

Manipulation of cross products, just like dot products, comes up regularly in calculus-based physics. The following are some important properties of cross products.

- $\vec{A} \times \vec{B} = -\vec{B} \times \vec{A}$

- $\vec{A} \times (\vec{B} + \vec{C}) = (\vec{A} \times \vec{B}) + (\vec{A} \times \vec{C})$

- $c(\vec{A} \times \vec{B}) = (c\vec{A}) \times \vec{B} = \vec{A} \times (c\vec{B})$

- $\dfrac{d}{dt}(\vec{A} \times \vec{B}) = \dfrac{d\vec{A}}{dt} \times \vec{B} + \vec{A} \times \dfrac{d\vec{B}}{dt}$

2.23 Q: Given the following vectors, perform the requested vector operations.

$$\vec{A} = <3m, 2m, -1m> \qquad \vec{B} = <-2m, -2m, 2m> \qquad \vec{C} = <1m, -3m, 0m>$$

A) $\vec{A} \cdot \vec{B} =$
B) $\vec{A} \cdot \vec{C} =$
C) $\vec{B} \cdot \vec{C} =$
D) $\vec{A} \times \vec{B} =$
E) $\vec{A} \times \vec{C} =$

F) $\vec{B} \times \vec{C} =$
G) $3\vec{A} \cdot (-\vec{B}) =$
H) $2\vec{B} \times \vec{C} =$
I) $(\vec{B} + \vec{C}) \cdot \vec{A} =$

2.23 A: A) -12 m²
B) -3 m²
C) 4 m²
D) <2 m², -4 m², -2 m²>
E) <-3 m², -1 m², -11 m²>

F) <6 m², 2 m², 8 m²>
G) 36 m²
H) <12 m², 4 m², 16 m²>
I) -15 m²

Graphing

Analyzing data is a fundamental skill in physics, and graphing is a fantastic way to visually evaluate data to recognize trends and relationships. Most graphs in physics show the dependent variable along the y-axis, and the independent variable along the x-axis, so that if a function is plotted such that y=f(x), x-values are plotted on the x-axis, and y-values, the output of the function f, are plotted on the y-axis.

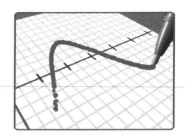

Dependent variables are variables that change in response to changes in the independent variables. **Independent variables** are typically the variables the experimenter is adjusting, and measuring the output in terms of dependent variables.

In creating a graph for scientific analysis, the following steps may be valuable.

1. Identify which variable is the dependent variable, and which variable is the independent variable.
2. Choose scales for your axes which make your graph as large as conveniently possible. Plot the scale for the independent variable on the x-axis, and the scale for the dependent variable on the y-axis. Label the axes, including units.
3. Plot the data points. Include the origin as a data point only if it was actually recorded as a data point, or if common sense indicates it is a data point.
4. Draw a best-fit line or smooth curve through your data points, with roughly the same number of points above and below the line. Do not connect the individual data points. Note that your best-fit line may touch all, none, or some portion of your data points.
5. Label your graph, noting that most graphs are titled in terms of the independent variable vs. dependent variable. For example, a graph of velocity on the y-axis vs. time on the x-axis could be labeled "Velocity vs. Time."

Once your graph is complete, an examination of the graph can provide insight into the relationship between the dependent and independent variables. If the graph is linear, you can analyze the slope and y-intercept of the graph to determine the equation of the line and therefore a relationship between the variables. If the graph is non-linear, you can attempt to linearize the graph by manipulating either the dependent or independent variables as indicated in the illustration below. You may then re-plot the data, and having obtained a linear relationship, analyze that line to determine the relationship between your dependent and independent variables.

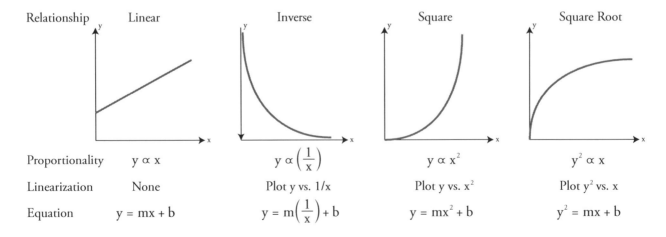

Relationship	Linear	Inverse	Square	Square Root
Proportionality	$y \propto x$	$y \propto \left(\frac{1}{x}\right)$	$y \propto x^2$	$y^2 \propto x$
Linearization	None	Plot y vs. 1/x	Plot y vs. x^2	Plot y^2 vs. x
Equation	$y = mx + b$	$y = m\left(\frac{1}{x}\right) + b$	$y = mx^2 + b$	$y^2 = mx + b$

2.24 Q: Given the following data describing the distance a skydiver falls from an airplane as a function of time, graph the data and determine a best mathematical fit of the variables.

Trial	1	2	3	4	5
Time (s)	0	1	2	3	4
Distance (m)	0	4.7	20	45	77

2.24 A: First plot the data as presented.

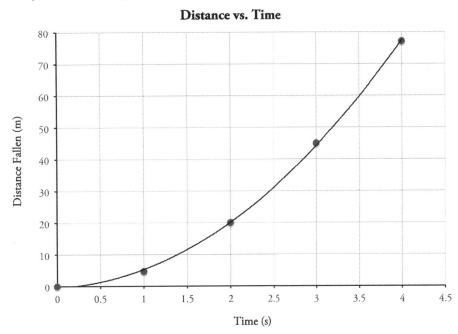

The best-fit line indicates this could be a square relationship, so you can re-plot the data, this time plotting distance fallen on the y-axis, and the square of time on the x-axis.

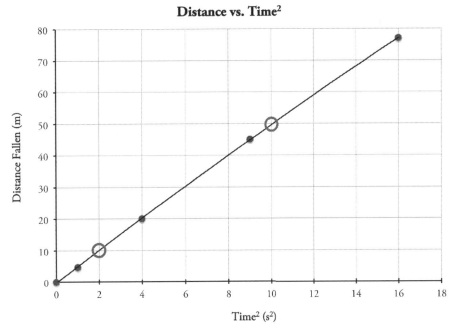

The best-fit line is now linear, indicating you can use the equation of a line in y=mx+b form to fit the function. Utilizing two points on the best fit line, not necessarily data points, you can find the slope of the best-fit line. Circle those points used in your calculations.

$$m = slope = \frac{rise}{run} = \frac{y_2 - y_1}{x_2 - x_1} = \frac{50m - 10m}{10s^2 - 2s^2} = 5\frac{m}{s^2}$$

From the graph, you can also determine the y-intercept is zero. Use this data to customize your equation of a line to determine the relationship between your variables.

$$y = mx + b \xrightarrow[x=t]{y=d} d = mt + b \xrightarrow[b=0]{m=5\frac{m}{s^2}} d = \left(5\frac{m}{s^2}\right)t^2$$

Calculus

And now for the topic that instills fear in the hearts of the bravest of souls: calculus. But guess what? You probably already know some calculus, and the level of calculus used in this course is fairly basic. Leaving the deep derivations and background information to math courses, this book will focus on problem solving. Having said that, if the calculus starts to trip you up, there's absolutely nothing wrong with diverting some time to review calculus principles using textbooks, review books, Internet tutorials, or even Aunt Janice (assuming Aunt Janice knows a bit of calculus and is willing to spend some time with you).

In its simplest form, calculus can be broken up into two activities, which turn out to be very closely related. Differential, or derivative calculus, involves finding the slope of a line, something you've probably been doing for years now. More precisely, it is finding how much one variable changes as you change another variable, a rate of change. For example, speed is the rate at which an object's distance traveled changes with respect to time. Speed, therefore, can be thought of as as the derivative of distance with respect to time. If you were to plot distance traveled on the y-axis as a function of time elapsed on the x-axis, the speed of the object at any specific time would be the slope of the line tangent to the graph at that time.

Integral calculus, on the other hand, involves finding the area under a graph, also something you've likely had some experience with. It is, in essence, taking the anti-derivative, or starting with the slope of a function, and determining the original function. Building upon our previous analogy, if speed is the derivative of distance with respect to time, distance is the integral of speed with respect to time.

What is new to these activities is calculating slopes and areas from their functional representations instead of relying solely on geometric solutions. Calculus allows you to accurately and efficiently find slopes and areas of functions, which mirror many of the phenomena observed in the natural world, and therefore are extremely valuable for modeling the world around us using physics class principles.

Let's start by looking at differential calculus, the calculus of slopes. Let's define a function f, which takes as its input some value x, and outputs some value y. It's almost like f is a magic black box, and when you put something into it, which we'll call x, the output is something else, y. The mathematical relationship that describes what you get out (your y) based on what you put in (your x) is the function f.

$$f(x)$$

$$x \qquad y = f(x) \qquad y$$

If you draw a graph of the output on the y-axis, and the input on the x-axis, you might obtain something similar to the plot below.

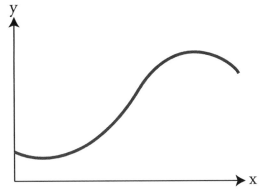

You may note that the slope of the graph above is not constant. Therefore, in order to take the slope of the graph, you have to specify at which point on the graph you want the slope. Since you can't truly take the slope of a curved line, you'll have to make an approximation. Using two points on the graph, you can connect those points with a line to approximate the slope. As the points get closer and closer to each other, you can draw a line tangent to the graph that becomes a better and better approximation to the slope of the function at that point.

> **Activity:** Draw a curved plot on a sheet of paper, take a piece of uncooked spaghetti and attempt to overlay the spaghetti on the function. Now break the uncooked spaghetti and repeat the effort. Note that as you use smaller and smaller pieces of uncooked spaghetti, your spaghetti graph gets closer and closer to the actual function. This is similar to what happens as you analyze tangent lines using points on the graph that get closer and closer to each other. Differential calculus is the mathematical process of finding the tangent to the curve by using points that get closer and closer to each other until they are "infinitesimally" close to each other.

Calculus Notation

And now for a quick aside on notation. For many, the notation of calculus is the trickiest part. Understanding this notation up front can greatly simplify future work, and alleviate many concerns. The first notation you are going to encounter is the **differential**, using the symbol *d*. All this symbol means is a tiny bit of something, so tiny that the amount approaches 0. Therefore, the notation *dx* would be read as "a tiny amount of x." The ratio *dx/dt* is the tiny amount that x changes for some tiny amount of change in t, and is read out loud as "the derivative of x with respect to t." The underlying concepts are straightforward, it's just the notation and verbalization that gets tricky.

Alternately, you could write this in functional form. The expression *y=f(x)* describes a function *f*, which takes an input variable *x* and places the output in variable y. You could plot this graphically with y values on the y-axis, and x-values on the x-axis. The derivative of y with respect to x, written *dy/dx*, describes the rate at which the y-value changes for a given change in the x-value. It also describes the slope of the graph at any given point as a function of x, and can also be written as $y' = f'(x) = \frac{dy}{dx}$, where the prime symbol indicates a derivative. If you were to take the derivative of the derivative, you obtain a second derivative, which can be written in a variety of ways as well, such as $y'' = f''(x) = \frac{d^2 y}{dx^2}$, where $\frac{d^2 y}{dx^2}$ does not indicate the squaring of variables, but rather indicates you are taking the second derivative of y with the respect to x.

Leaving further derivations to calculus classes, you should note that there are various types of functions for which determining the slope of the tangent at any point is fairly straightforward. In many cases, you'll find that the slope of the line at any point is given by another function. This function is what is known as the derivative of the initial function.

Looking at notation on the integral side, the integral symbol ∫ in its essence means add up all the pieces of whatever follows the symbol. The expression ∫dt would therefore mean "add up all the little bits of dt," where dt is the differential of time, so putting it all together, ∫dt means "add up all the little bits of time."

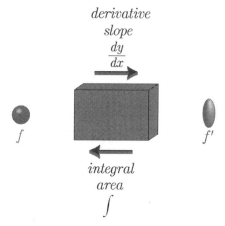

Like many things in physics, derivatives and integrals are considerably more complex than actually using them. So with that, let's introduce some common derivatives and integrals and utilize examples to get a better handle on the concepts.

Derivatives of Constants

The derivative of a constant is zero. The plot of a constant is a straight horizontal line, so its slope at any point is zero, its rate of change is zero, and its derivative is zero. Written mathematically, if C is a constant:

$$\frac{d}{dx}(C) = 0$$

This would be read as "the derivative of C with respect to x equals zero."

Derivatives of Polynomials

The derivative of a first-order polynomial is a constant. The derivative of a second-order polynomial is a first polynomial. The derivative of a third-order polynomial is a second-order polynomial. The specific formula for polynomial derivatives is given by:

$$\frac{d}{dx}(x^n) = nx^{(n-1)}$$

2.25 Q: Find the following derivatives:
A) $y=x^7$
B) $u=x^{6a}$

2.25 A: A) $\dfrac{dy}{dx} = \dfrac{d}{dx}(x^7) = 7x^6$

B) $\dfrac{du}{dx} = \dfrac{d}{dx}(x^{6a}) = 6ax^{6a-1}$

Rules for Derivatives

Let us now discuss some basic rules for derivatives that will allow you to work with more complex functions. First, the derivative of a constant multiplied by a function is equal to the constant multiplied by the derivative of the function. Or, put more simply, you can pull constants out of the derivative sign.

2.26 Q: Find the following derivatives with respect to x:
A) $y=4x^3$
B) $f(x)=12x^2$
C) $3\left(\sqrt[3]{x^a}\right)$

2.26 A: A) $\dfrac{dy}{dx} = \dfrac{d}{dx}(4x^3) = 12x^2$

B) $f' = 24x$ (Note that this is actually the second derivative of part A, $4x^3$.)

C) $\dfrac{d}{dx}3\left(\sqrt[3]{x^a}\right) = 3\dfrac{d}{dx}\left(x^{\frac{a}{3}}\right) = 3\left(\frac{a}{3}x^{\frac{a}{3}-1}\right) = ax^{\frac{a}{3}-1}$

What if you want to take the derivative of a sum or a difference? This is actually fairly straightforward. The derivative of a sum is the sum of the derivatives (and likewise for differences). Put more simply, you can differentiate the different terms in a sum or difference separately.

2.27 Q: Find the following derivatives:
A) $\dfrac{d}{dx}\left(-3x^4 + 2x\right)$

B) $\dfrac{d}{dz}\left(5z^{2.5} - 2\right) = ?$

2.27 A: A) $y' = -12x^3 + 2$

B) $\dfrac{d}{dz}\left(5z^{2.5} - 2\right) = 12.5z^{1.5} - 0 = 12.5z^{1.5}$

You could also be asked to determine the derivative of a product or quotient. Though the process isn't quite as straightforward, it is still quite reasonable. The derivative of a product is found by taking the derivative of the first function multiplied by the second function, and adding that to the first function multiplied by the derivative of the second function. If the two functions are u and v, this may be written as $(uv)' = u'v + uv'$.

2.28 Q: Find the following derivatives with respect to x using the product rule:
A) $y = \sqrt{x^3}\,(x^2 - 3x)$
B) $f(x) = (4x^3 + x)(6 - 3x)$

2.28 A: A) Define u and v as follows: $u = \sqrt{x^3} = x^{\frac{3}{2}}, \; v = x^2 - 3x$

Use the product rule: $\dfrac{d}{dx}(uv) = v\dfrac{du}{dx} + u\dfrac{dv}{dx} \rightarrow y' = (x^2 - 3x)(1.5x^{\frac{1}{2}}) + x^{\frac{3}{2}}(2x - 3)$

Simplify: $y' = 1.5x^{\frac{5}{2}} - 4.5x^{\frac{3}{2}} + 2x^{\frac{5}{2}} - 3x^{\frac{3}{2}} = 3.5x^{\frac{5}{2}} - 7.5x^{\frac{3}{2}} = x^{\frac{3}{2}}(3.5x - 7.5)$

B) Define u and v as follows: $u = 4x^3 + x, \; v = 6 - 3x$

Use the product rule: $f'(x) = (6 - 3x)(12x^2 + 1) + (4x^3 + x)(-3)$

Multiply: $f'(x) = 72x^2 + 6 - 36x^3 - 3x - 12x^3 - 3x$

Simplify: $f'(x) = -48x^3 + 72x^2 - 6x + 6 = 6(-8x^3 + 12x^2 - x + 1)$

The formula for the derivative of the quotient is given by: $\left(\dfrac{u}{v}\right)' = \dfrac{u'v - uv'}{v^2}$.

A method of remembering this formula is to define the function f as "hi" over "ho": $f(x) = \dfrac{hi}{ho}$. The derivative of f, then, which is the derivative of hi/ho, can be read as ho dee hi minus hi dee ho all over ho ho.

2.29 Q: Find the following derivatives with respect to x using the quotient rule:
A) $y = \dfrac{2x + 5}{x - 2}$

B) $f(x) = \dfrac{3\sqrt[3]{x}}{x^2 - 1}$

2.29 A: A) Define u and v as: $u = 2x + 5, \; v = x - 2$

Use the quotient rule: $y' = \left(\dfrac{u}{v}\right)' = \dfrac{u'v - uv'}{v^2} = \dfrac{2(x - 2) - (2x + 5)(1)}{(x - 2)^2}$

Simplify: $y' = \dfrac{2(x - 2) - (2x + 5)(1)}{(x - 2)^2} = \dfrac{-9}{(x - 2)^2}$

B) Define u and v as: $u = 3\sqrt[3]{x}, \; v = x^2 - 1$

Use the quotient rule: $f'(x) = \dfrac{u'v - uv'}{v^2} = \dfrac{x^{-\frac{2}{3}}(x^2 - 1) - (3x^{\frac{1}{3}})(2x)}{(x^2 - 1)^2}$

Simplify: $f'(x) = \dfrac{x^{\frac{4}{3}} - x^{-\frac{2}{3}} - 6x^{\frac{4}{3}}}{(x^2 - 1)^2} = \dfrac{x^2 - 1 - 6x^2}{x^{\frac{2}{3}}(x^2 - 1)^2} = \dfrac{-5x^2 - 1}{x^{\frac{2}{3}}(x^2 - 1)^2}$

Chapter 2: Background Skills

Derivatives of Trig Functions

You can also find the derivative of trigonometric functions. Though you will likely derive some (or all) of these in your calculus class, just the formulas are provided here. Note that in AP Physics C you will deal mostly with the sine, cosine, and tangent functions, though the derivatives of all six trigonometric functions are provided here for completeness.

$$\frac{d}{dx}(\sin(ax)) = a\cos(ax) \qquad\qquad \frac{d}{dx}(\cos(ax)) = -a\sin(ax)$$

$$\frac{d}{dx}(\tan(ax)) = a\sec^2(ax) \qquad\qquad \frac{d}{dx}(\cot(ax)) = -a\csc^2(ax)$$

$$\frac{d}{dx}(\sec(ax)) = a\sec(ax)\tan(ax) \qquad\qquad \frac{d}{dx}(\csc(ax)) = -a\csc(ax)\cot(ax)$$

2.30 Q: Find the following derivatives with respect to the given input variable:
A) y(x)=3sin(3x)
B) f(x)=cos(-2x)
C) y(x)=cos²(2x)-sin²(2x)
D) y(a)=4sin(3a)cos(3a)
E) y(θ)=sin(θ)/cos(θ)
F) f(x)=sin(3x)/cos(x)

2.30 A: A) $y' = 9\cos(3x)$

B) $f'(x) = 2\sin(-2x)$

C) Note that this can be simplified before differentiating using the double angle formula, which will greatly simplify the differentiation: $\cos^2(2x) - \sin^2(2x) = \cos(4x)$. Now this becomes a straightforward cosine derivative. $y' = -4\sin(4x)$

D) This can be simplified using the sine double angle formula, so that y=2sin(6a). This now becomes a straightforward sine derivative. $y' = 12\cos(6a)$

E) This can be simplified by recognizing this is the definition of the tangent, so y=tan(θ). Using the derivative of a tangent formula: $y' = \sec^2(\theta)$.

F) You can tackle this one using the quotient formula. Define u and v as:
$u = \sin(3x), \quad v = \cos(x)$.

Use the quotient rule: $f'(x) = \left(\frac{u}{v}\right)' = \frac{u'v - uv'}{v^2} = \frac{3\cos(3x)\cos(x) + \sin(3x)\sin(x)}{\cos^2(x)}$

Apply the sum-difference formula for cosines: $f' = \frac{3\cos(3x)\cos(x) + \sin(3x)\sin(x)}{\cos^2(x)}$

Of course, if you feel the need, you could simplify this further. Not feeling the need personally, allow me to move on.

Derivatives of Exponential and Logarithmic Functions

Next we'll take a look at the natural exponential and logarithmic functions, as they are quite common in physics, and fairly easy to differentiate. Beginning with the natural exponential:

$$\frac{d}{dx}(e^{ax}) = ae^{ax}$$

In essence, the derivative of the natural exponential is the natural exponential! Closely related, though perhaps not in an obvious manner (at least not yet), is the natural logarithm function.

$$\frac{d}{dx}(\ln ax) = \frac{1}{x}$$

Let's try a few.

2.31 Q: Find the following derivatives:
A) y=e^{3x}
B) f(x)=ln(6x)

2.31 A: A) $y' = 3e^{3x}$

B) $f'(x) = \frac{1}{x}$

The Chain Rule

One last major rule to cover, and this one is extremely useful, as it allows you to find the derivative of a function that has as its argument another function. Imagine, for example, a function such as $y = \sin(3x^2)$. You could write this in the form y=f(g(x)), where f refers to the sine function, and g refers to the polynomial. The derivative of y can be found using the chain rule:

$$y' = f'(g(x)) \times g'(x)$$

Or, as it is written on the formula sheet:

$$\frac{df}{dx} = \frac{df}{du}\frac{du}{dx}$$

Regardless of your preferred form, the calculation works the same way. Let's work through the first example slowly.

2.32 Q: Find the derivative of y=sin(3x²).

2.32 A: First let's define the pieces of this function.

$$f(x) = \sin(x) \qquad g(x) = 3x^2$$
$$f'(x) = \cos(x) \qquad g'(x) = 6x$$
Now, utilize the chain rule to find the derivative of y.
$$y' = f'(g(x)) \times g'(x) = \cos(3x^2) \times 6x = 6x\cos(3x^2)$$

There, that wasn't so bad. Try some more to make sure you've got it, as this is a rule you'll use throughout the entire course.

2.33 Q: Find the following derivatives:
A) $y = \cos\left(5x^3 - 2x\right)$

B) $f(x) = \left[3x^2 + \cos(x)\right]^3$

C) $I(t) = 8e^{-\frac{t}{4}}$

D) $y = \ln\left(6x^4\right)$

2.33 A: A) $y' = -\sin\left(5x^3 - 2x\right)\left(15x^2 - 2\right) = \left(2 - 15x^2\right)\sin\left(5x^3 - 2x\right)$

B) $f'(x) = 3\left[3x^2 + \cos(x)\right]^2\left(6x - \sin x\right)$

C) $I'(t) = 8e^{-\frac{t}{4}}\left(-\frac{1}{4}\right) = -2e^{-\frac{t}{4}}$

D) $y' = \frac{1}{6x^4}\left(24x^3\right) = \frac{4}{x}$

Integration

The other half of calculus, integral calculus, is closely related to differential calculus. When you took the derivative of a function, you found the slope of the original function. The integral allows you to go the other way, and is often times called the "anti-derivative." As an example, if you started with a function y=3x², its derivative is y'=6x. Looking at this from the opposite perspective, if you start with a function y=6x, the integral of 6 is 3x² (plus some constant, which we'll talk about shortly).

As a brief reminder about notation, the integral sign, "∫", means add up all the little pieces of something, even if there are an infinite number of infinitessimally small little pieces. If you have some function y and you chop it up into an infinite number of infinitessimally small pieces, which we call dy, it would be impossible to add them all up physically. Mathematically, however, if you were to sum up all those infinitessimally small pieces dy, you would obtain the total you started with, y. Written mathematically:

$$\int dy = y$$

Further, if the derivative of a function gives you the slope of that function, the integral of a function gives you the area under the function. Referring to the graph below showing a function y=f(x):

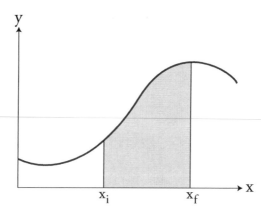

You can find the area under the graph between the initial x (x_i) and final x (x_f) values by integration. Mathematically, you do this by breaking up the function into lots of thin little rectangles of width dx and height equal to the function's value at that x. If you were to add up the area of all those rectangles, you'd approximate the area under the curve. To get more exact, you make the rectangles skinnier, adding more of them. If you could make them infinitessimally small, you'd have an infinite amount of rectangles. Add all those up by integrating, and you get the exact area!

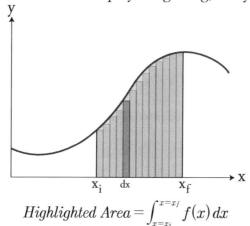

$$Highlighted\ Area = \int_{x=x_i}^{x=x_f} f(x)\,dx$$

Writing this mathematically, you'll note the addition of some further terms to the integral sign. The bottom term attached to the integral sign indicates the starting point of our integration, and the top term indicates the ending point. These are referred to as the **limits of integration**.

Integrals of Polynomials

You can probably figure out the formula for the integral of a polynomial from the formula of the derivative of a polynomial:

$$\int x^n\,dx = \frac{1}{n+1}x^{n+1},\ n \neq -1$$

Note that you cannot use this formula for an n value of -1, as you would end up with a zero in the denominator. If you do come across an integration in this form with an n value of -1, rewrite the function as dx/x, which will allow you to use a logarithm integral formula, introduced shortly.

There is one problem with this formula, and it's a problem that is inherent to all integral formulas. If you recall, the derivative of a constant is zero. The derivative of six is zero, the derivative of 60,000 is zero. So when you're going the other way, starting with the derivative and attempting to backtrack, or integrate, to the original function, how can you account for that unknown constant? Not knowing what the constant is, we'll assign it to a variable C, known as the constant of integration. We typically don't know what C's value is, but we know it can be any constant. If you want to find out more specifically what C is, you'll need a bit more information from the problem, typically referred to as boundary conditions. These boundary conditions can be information about the original function's value at a specific point, or they can be the limits of integration highlighting beginning and ending points.

Integration problems in which the limits of integration aren't given are known as **indefinite integrals**. Integration problems in which the limits of integration are known are called **definite integrals**. Like derivatives, you can pull constants out of the integral sign. Also similar to derivatives, the integral of a sum is the sum of the integrals.

Let's start with a few polynomial indefinite integrals.

2.34 Q: Find the following integrals:

A) $\int (x^3 - 2x + 9)\, dx$

B) $\int 3dy$

C) $\int \frac{6x^4 + 2x^3 - 4x}{2x}\, dx$

2.34 A: A) First realize that the integral of a sum is the sum of the integrals:
$$\int (x^3 - 2x + 9)\, dx = \int x^3\, dx - \int 2x dx + \int 9 dx$$

Next, you can pull the constants out of the integral sign.
$$\int x^3\, dx - \int 2x dx + \int 9 dx = \int x^3\, dx - 2\int x dx + 9\int dx$$

Then, integrate the polynomial terms.
$$\int x^3\, dx - 2\int x dx + 9\int dx = \left(\frac{x^4}{4} + C\right) - \left(2\frac{x^2}{2} + C\right) + (9x + C)$$

Finally, you have three unknown constants. Recognizing that the sum of three unknown constants is still an unknown constant, you can re-write your equation to obtain the final answer.
$$\left(\frac{x^4}{4} + C\right) - \left(2\frac{x^2}{2} + C\right) + (9x + C) = \frac{x^4}{4} - x^2 + 9x + C$$

B) $\int 3dy = 3\int dy = 3y + C$

C) Instead of looking at a quotient rule for integrals, you can simplify this problem by dividing the numerator by 2x.
$$\int \frac{6x^4 + 2x^3 - 4x}{2x}\, dx = \int (3x^3 + x^2 - 2)\, dx$$

Next, integrate the polynomial.
$$\int (3x^3 + x^2 - 2)\, dx = 3\frac{x^4}{4} + \frac{x^3}{3} - 2x + C$$

2.35 Q: If y'=6t, determine the function y, knowing the boundary condition of y at time t=0 is equal to -2.

2.35 A: $y = \int 6t\,dt = 6\int t\,dt = 6\frac{t^2}{2} + C = 3t^2 + C$

Apply the boundary condition that y(0)=-2 to find the value of C.

$y(0) = -2 \xrightarrow{y=3t^2+C} 3t^2 + C = -2 \xrightarrow{t=0} C = -2$

Finally, knowing C=-2, you can substitute back into equation for y.

$y = 3t^2 + C \xrightarrow{C=-2} y = 3t^2 - 2$

Now you can try a few definite integrals. The process for solving definite integrals is perhaps best explained by demonstration.

2.36 Q: Find the following integral: $\int_{x=2}^{x=4} (x^3 - 2x + 9)\,dx$

2.36 A: First solve the indefinite integral.

$$\int_{x=2}^{x=4} (x^3 - 2x + 9)\,dx = \left.\left(\frac{x^4}{4} - x^2 + 9x\right)\right|_{x=2}^{x=4}$$

Note the new symbology at the right side of the answer. The vertical line at the right indicates that the term to the left is to be evaluated from x=2 to x=4. You take the top value and substitute it into the term, then subtract the term with the bottom value substituted in place of the variable.

$$\left.\left(\frac{x^4}{4} - x^2 + 9x\right)\right|_{x=2}^{x=4} = \left(\frac{4^4}{4} - 4^2 + 9(4)\right) - \left(\frac{2^4}{4} - 2^2 + 9(2)\right)$$

Obtain the final answer by performing the calculation.

$$\left(\frac{4^4}{4} - 4^2 + 9(4)\right) - \left(\frac{2^4}{4} - 2^2 + 9(2)\right) = 66$$

Now that you've seen the notation, practice a few to test your understanding.

2.37 Q: Evaluate the following

A) $\int_{t=0}^{3} 6t^2\,dt$

B) Find the area under the graph of y=x² from x = 0 to x = π.

2.37 A: Evaluate the following

A) $\int_{t=0}^{3} 6t^2\,dt = 6\int_{t=0}^{3} t^2\,dt = 6\left.\frac{t^3}{3}\right|_{t=0}^{3} = 6\frac{3^3}{3} - 6\frac{0^3}{3} = 54$

B) $\int_{x=0}^{\pi} x^2\,dx = \left.\frac{x^3}{3}\right|_0^{\pi} = \frac{\pi^3}{3} - 0 = \frac{\pi^3}{3}$

Integrals of Trigonometric Functions

Just as you could take the derivative of trigonometric functions, you can also integrate trig functions.

$$\int \sin u\, du = -\cos u + C \qquad\qquad \int \cos u\, du = \sin u + C$$

$$\int \sec^2 u\, du = \tan u + C \qquad\qquad \int \sec u \tan u\, du = \sec u + C$$

$$\int \csc^2 u\, du = -\cot u + C \qquad\qquad \int \csc u \cot u\, du = -\csc u + C$$

2.38 Q: Find the area under the graph of $y = \sec^2(3\theta)$ between 0 and $\pi/16$.

2.38 A: First integrate: $\int_{\theta=0}^{\pi/16} \sec^2(3\theta)\, d\theta$. You'll notice that this isn't really in the form of the formula for the integral of the square of the secant. To fit the form, our u, equal to 3θ, must be followed by du, which would be $3d\theta$. To make this fit the form, then, multiply by 3 inside the integral sign. To maintain the value of the term, however, we'll have to multiply by 1/3 outside the integral sign.

$$\int y\, dy = \int_{\theta=0}^{\pi/16} \sec^2(3\theta)\, d\theta \xrightarrow[du=3d\theta]{u=3\theta} \int y\, dy = \frac{1}{3}\int_{\theta=0}^{\pi/16} \sec^2(3\theta)\, 3d\theta$$

Now, you can integrate using the trig formula above.

$$\frac{1}{3}\int_{\theta=0}^{\pi/16} \sec^2(3\theta)\, 3d\theta = \tan(3\theta)\,|_{0}^{\pi/16} = \frac{1}{3}\left(\tan\left(\frac{3\pi}{16}\right) - \tan(0)\right) = 0.223$$

2.39 Q: Find the following integrals
A) y'=9cos(3x)
B) f'(x)=2sin(-2x)
C) y'=-4sin(4x)
D) f'(x)=12cos(6x), f(0)=0

2.39 A: A) $y = \int 9\cos(3x)\, dx = 9\int \cos(3x)\, dx \xrightarrow[du=3dx]{u=3x} y = \frac{9}{3}\int \cos(3x)\, 3dx = 3\sin(3x) + C$

B) $f(x) = \int f'(x)\, dx = \int 2\sin(-2x)\, dx = -\int \sin(-2x)\, 2dx = \cos(2x) + C$

C) $y = \int -4\sin(4x)\, dx = -4\int \sin(4x)\, dx = -\frac{4}{4}\int \sin(4x)\, 4dx = \cos(4x) + C$

D) $f(x) = \int f'(x)\, dx = \int 12\cos(6x)\, dx = \frac{12}{6}\int \cos(6x)\, 6dx = 2\sin(6x) + C$

Next, apply the boundary conditions to find C.
$$f(0) = 0 \xrightarrow{f(x)=2\sin(6x)+C} 2\sin(6x) + C = 0 \xrightarrow{x=0} C = 0$$

Finally, substitute in to obtain f(x).
$$f(x) = 2\sin(6x) + C \xrightarrow{C=0} f(x) = 2\sin(6x)$$

Derivatives of Exponential and Logarithmic Functions

Finally, we'll take a look at the exponential and logarithmic integrals.

$$\int e^u \, du = e^u + C \qquad\qquad \int \frac{du}{u} = \ln|u| + C$$

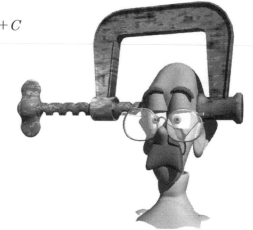

2.40 Q: Find the following integrals

A) $\int \frac{4}{2x - 3} \, dx$

B) $y' = 3e^x$

C) $f'(x) = (6 - 4e^{2x}) \, dx$

D) $\int \frac{4x^3}{2x^4 + 3} \, dx$

2.40 A: A) First you must make the problem fit the formula.

$$y = \int \frac{4}{2x - 3} \, dx \xrightarrow[du = 2dx]{u = 2x - 3} y = 2 \int \frac{2 \, dx}{2x - 3}$$

As this is now in the form of du/u, you can apply the integral formula for the natural log:

$$y = 2 \int \frac{2 \, dx}{2x - 3} \xrightarrow{\int \frac{du}{u} = \ln|u| + C} y = 2\ln|2x - 3| + C$$

B) $\displaystyle\int 3e^x \, dx = 3 \int e^x \, dx = 3e^x + C$

C) First break the integral up into its constituent parts and pull the constants out of the integral sign.

$$f(x) = \int f'(x) \, dx = \int (6 - 4e^{2x}) \, dx = \int 6 \, dx - \int 4e^{2x} \, dx = 6\int dx - 4\int e^{2x} \, dx$$

Next, you can make the exponential function fit the form of the integral formula.

$$f(x) = 6\int dx - 4\int e^{2x} \, dx = 6\int dx - \frac{4}{2}\int e^{2x} 2 \, dx = 6\int dx - 2\int e^{2x} 2 \, dx$$

Finally, integrate the terms. Don't forget, two unknown constants added together gives you an unknown constant.

$$f(x) = 6\int dx - 2\int e^{2x} 2 \, dx \xrightarrow{\int e^u \, du = e^u + C} f(x) = 6x + C - 2e^{2x} + C = 6x - 2e^{2x} + C$$

D) $\displaystyle y = \int \frac{4x^3}{2x^4 + 3} \, dx \xrightarrow[du = 8x^3 dx]{u = 2x^4 + 3} y = \frac{1}{2}\int \frac{8x^3 \, dx}{2x^4 + 3} = \frac{1}{2}\ln|2x^4 + 3| + C$

Of course, there is considerably more to calculus than is covered here, but hopefully this will give you a reasonable start, while leaving the details and complexities to your calculus class. Let's finish off with a couple more examples and an application problem that will be further explained in the next chapter.

2.41 Q: Integrate the following functions.

A) $\int 3x^2\, dx =$

B) $\int 2\cos(6x)\, dx =$

C) $\int e^{4x}\, dx =$

D) $\int_{x=0}^{x=4} 2x\, dx =$

2.41 A: A) $x^3 + C$

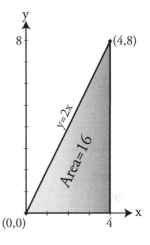

B) $\int 2\cos(6x)\, dx = \frac{2}{6}\int \cos(6x)\, 6dx = \frac{\sin(6x)}{3} + C$

C) $\int e^{4x}\, dx = \frac{1}{4}\int e^{4x}\, 4dx = \frac{e^{4x}}{4} + C$

D) $\int_{x=0}^{x=4} 2x\, dx = x^2 \big|_{x=0}^{x=4} = 4^2 - 0^2 = 16$

Note that the answer for D is the same as the answer you'd have arrived at if you had taken the area under the graph of y=2x from x=0 to x=4, as shown at right.

2.42 Q: The velocity of a particle as a function of time is given by the equation $v(t) = 3t^2$. The particle starts at position 0 at time 0.

A) Find the slope of the velocity-time graph as a function of time. This gives you the particle's acceleration function.

B) Find the area under the velocity-time graph as a function of time. This gives you the particle's position function.

2.42 A: A) $a(t) = slope = v'(t) = \frac{d}{dt}(3t^2) = 6t$

B) $r(t) = \int v(t)\, dt = \int 3t^2\, dt = t^3 + C \xrightarrow[r(0)=0]{B.C.} 0^3 + C = 0 \xrightarrow{C=0} r(t) = t^3$

Author's Note: The solutions to these problems as written indicate a dimension problem. For example, r(t)=t³ would seem to indicate that position is given in units of s³, which of course doesn't make sense. What actually occurs is during the integration we have neglected units by dropping a constant of 1 m/s³, which would work out to r(t)=t³×1 m/s³ if we followed through more precisely. However, keeping track of units while taking derivatives and integrals is often times more complex than solving the problem itself, therefore it is common practice to reduce complexity by neglecting units while working through problems. Written more completely with the correct dimensions, the answer to part (A) would be $a(t) = (6t)\left(1\tfrac{m}{s^3}\right)$, and (B)'s solution would be $r(t) = (t^3)\left(1\tfrac{m}{s^3}\right)$.

Chapter 2: Background Skills

Chapter 3: Kinematics

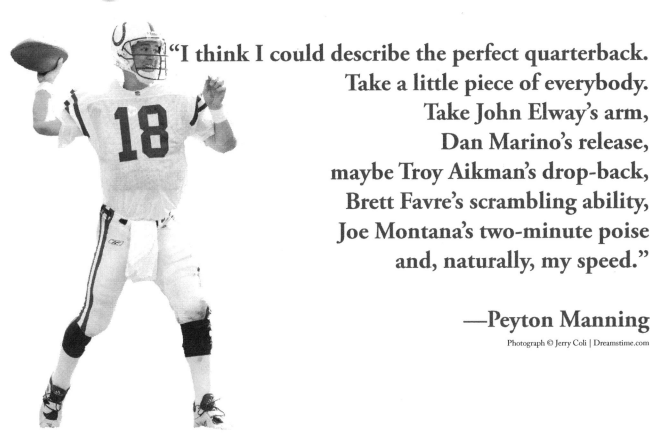

"I think I could describe the perfect quarterback. Take a little piece of everybody. Take John Elway's arm, Dan Marino's release, maybe Troy Aikman's drop-back, Brett Favre's scrambling ability, Joe Montana's two-minute poise and, naturally, my speed."

—Peyton Manning

Photograph © Jerry Coli | Dreamstime.com

Objectives

1. Calculate the kinetic energy of a moving object.

2. Explain the difference between distance and displacement, average velocity and instantaneous velocity, and average acceleration and instantaneous acceleration.

3. Utilize the general relationships among position, velocity, and acceleration for the motion of a particle to calculate unknown kinematic quantities.

4. Draw and interpret graphs of motion.

5. Write appropriate equations for catch-up problems and solve them for appropriate kinematic quantities.

6. Use kinematic equations to solve problems of motion with constant acceleration.

7. Recognize the independence of motion in the varying coordinate axes.

8. Describe the motion of a projectile as both a function of time and position.

9. Sketch the trajectory of a projectile.

10. Calculate the velocity of an object relative to various reference frames.

P hysics is all about energy in the universe, in all its various forms. Here on Earth, the source of all energy, directly or indirectly, is the conversion of mass into energy. We receive most of this energy from the sun. Solar power, wind power, hydroelectric power, fossil fuels, all can be traced back to the sun and the conversion of mass into energy. So where do you start in your study of the universe?

Theoretically, you could start by investigating any of these types of energy. In reality, however, by starting with energy of motion (also known as **kinetic energy**, K), you can develop a set of analytical problem solving skills from basic principles that will serve you well as you expand into the study of other types of energy.

For an object to have kinetic energy, it must be moving. Specifically, the kinetic energy of an object is equal to one half of the object's mass multiplied by the square of its velocity.

$$K = \frac{1}{2}mv^2$$

If kinetic energy is energy of motion, and **energy** is the ability or capacity to do work (moving an object), then you can think of kinetic energy as the ability or capacity of a moving object to move another object.

But what does it mean to be in motion? A moving object has a varying position. Its location changes as a function of time. So to understand kinetic energy, you'll need to better understand position and how position changes. This will lead into the first major unit, kinematics, from the Greek word kinein, meaning to move. Formally, kinematics is the branch of physics dealing with the description of an object's motion, leaving the study of the "why" of motion to the next major topic, dynamics.

Defining Kinematic Quantities

An object's location in space and time is known as its **position**. The vector from the origin of the coordinate system to the object's position is known as the position vector $\vec{r}$. For an object located on the x-axis at x=3, its position can be written as (3,0,0). The position vector for the object is therefore written as $\vec{r}(t) = <3,0,0> = 3\hat{i}$. If you are dealing with motion in one dimension, just along the x-axis, you could also write this as x=3.

Chapter 3: Kinematics

Objects that don't move aren't overly interesting from a physics perspective, at least not when you're studying motion. As an object moves as a function of time, its position changes, therefore the position vector is a function of time, written $\vec{r} = \vec{r}(t)$. Since the position vector has components in the x-, y-, and z-directions, you can also write this as $\vec{r}(t) = <x(t), y(t), z(t)> = x(t)\hat{i} + y(t)\hat{j} + z(t)\hat{k}$, where x(t), y(t), and z(t) are functions describing the x-, y-, and z-coordinates of the object as a function of time.

The change in the position vector from some initial time t_i to some final time t_f is called the **displacement** vector, written as $\Delta\vec{r}$, where $\Delta\vec{r} = \vec{r}(t_f) - \vec{r}(t_i)$. In three dimensions, the displacement vector can be determined from $\Delta\vec{r} = \vec{r}(t_f) - \vec{r}(t_i) = <x(t_f), y(t_f), z(t_f)> - <x(t_i), y(t_i), z(t_i)>$ or, written more simply, $\Delta\vec{r} = \vec{r}(t_f) - \vec{r}(t_i) = <\Delta x(t), \Delta y(t), \Delta z(t)>$. In one dimension, position is given by the x-coordinate, and displacement by Δx.

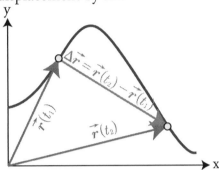

Summing up, an object's position refers to its location at any given point in time. The position vector ($\vec{r}$) begins at the origin and ends at the object's current location. Displacement ($\Delta\vec{r}$) is the vector which begins at the initial position of an object, and ends at the final position of the object. Finally, **distance**, as defined by physics, is a scalar. It has a magnitude, or size, only, and describes how far the object has traveled. The symbol for distance is Δs.

3.1 Q: On a sunny afternoon, a deer walks 1300 meters east to a creek for a drink. The deer then walks 500 meters west to the berry patch for dinner, before running 300 meters west when startled by a loud raccoon.
A) What distance did the deer travel?

B) What is the deer's displacement?

3.1 A: A) The deer traveled 1300m + 500m + 300m, for a total distance traveled of 2100m.

B) The deer's displacement was 500m east.

3.2 Q: A student on her way to school walks four blocks east, three blocks north, and another four blocks east, as shown in the diagram.

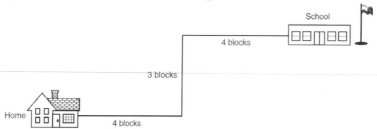

Compared to the distance she walks, the magnitude of her displacement from home to school is
(A) less
(B) greater
(C) the same

3.2 A: (A) The magnitude of displacement is always less than or equal to the distance traveled.

3.3 Q: A hiker walks 5 kilometers due north and then 7 kilometers due east. What is the magnitude of her resultant displacement? What total distance has she traveled?

3.3 A:

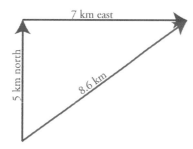

$$a^2 + b^2 = c \rightarrow c = \sqrt{a^2 + b^2} = \sqrt{(5km)^2 + (7km)^2} = 8.6km$$
The magnitude of the hiker's resultant displacement is 8.6 km.
The hiker's distance traveled is 12 kilometers.

Speed and Velocity

Knowing only an object's distance and displacement doesn't tell the whole story. Going back to the deer example, there's a significant difference in the picture of the deer's afternoon if the deer's travels occurred over 5 minutes (300 seconds) as opposed to over 50 minutes (3000 seconds).

How exactly does the picture change? In order to answer that question, you'll need to understand some new concepts: average speed and average velocity.

Average speed, given the symbol $\overline{v}$, is defined as distance traveled divided by time, and it tells you the rate at which an object's distance traveled changes. When applying the formula, you must make sure that Δs is used to represent distance traveled.

$$\overline{v} = \frac{\Delta s}{t}$$

Average velocity, on the other hand, is defined as the displacement divided by time, and is given the symbol $\vec{v}_{avg}$. Average velocity tells you the average rate of change of an object's position with respect to time.

$$\vec{v}_{avg} = \frac{\Delta \vec{r}}{\Delta t}$$

3.4 Q: A deer walks 1300 m east to a creek for a drink. The deer then walked 500 m west to the berry patch for dinner, before running 300 m west when startled by a loud raccoon. The entire trip took 600 seconds (10 minutes).
A) What is the deer's average speed?
B) What is the deer's average velocity?

3.4 A: A) $\bar{v} = \frac{s}{t} = \frac{2100m}{600s} = 3.5\,m/s$

B) $\vec{v}_{avg} = \frac{\Delta \vec{r}}{\Delta t} = \frac{\vec{r}_f - \vec{r}_i}{600s} = \frac{(500m)\,\hat{i}}{600s} = 0.83\,m/s\,\hat{i}$

3.5 Q: Find the average velocity between t=1 and t=5s of the object whose position as a function of time is shown in the graph below.

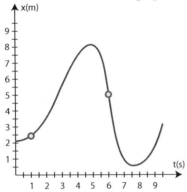

3.5 A: This can be calculated numerically and graphically. Starting with the numeric method:

$$\vec{v}_{avg} = \frac{\Delta \vec{r}}{\Delta t} = \frac{\vec{r}_f - \vec{r}_i}{\Delta t} = \frac{(5m)\,\hat{i} - (2.5m)\,\hat{i}}{6s - 1s} = (0.5\,m/s)\,\hat{i}$$

Graphically, you can calculate this by taking the slope of the line created by connecting the two points of interest, as shown below.

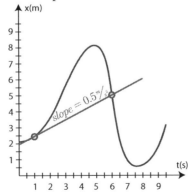

Acceleration

So you're starting to get a pretty good understanding of motion. But what would the world be like if velocity never changed? Objects at rest would remain at rest. Objects in motion would remain in motion at a constant speed and direction. Kinetic energy would never change. It'd make for a pretty boring world. Thankfully, velocity can change, and this change in velocity is known as an **acceleration**.

More accurately, acceleration is the rate of change of velocity with respect to time. You can write this as:

$$\vec{a}_{avg} = \frac{\Delta \vec{v}}{\Delta t}$$

This formula indicates that the change in velocity divided by the time interval gives you the average acceleration. It is also the average slope of the velocity-time graph. Much like displacement and velocity, acceleration is a vector—it has a direction. Further, the units of acceleration are meters per second per second, or $^m\!/_{s^2}$. Although it sounds complicated, all the units mean is that velocity changes at the rate of one meter per second, every second. So an object starting at rest and accelerating at 2 m/s² would be moving at 2 m/s after one second, 4 m/s after two seconds, 6 m/s after three seconds, and so on.

3.6 Q: Monty the Monkey accelerates in a rocket car from rest to a velocity of 9 m/s in a time span of 3 seconds. Calculate Monty's average acceleration.

3.6 A: $\vec{a}_{avg} = \frac{\Delta \vec{v}}{\Delta t} = \frac{\vec{v}_f - \vec{v}_i}{\Delta t} = \frac{9^m\!/_s - 0^m\!/_s}{3s} = 3^m\!/_{s^2}$

Note that since no direction is specified in the problem, no direction is given in the answer. It is assumed that the acceleration is in the forward direction based on the problem's context.

Instantaneous Velocity and Acceleration

You can determine the instantaneous velocity, the rate of change of an object's position at a specific instant in time, by taking the average velocity over an infinitessimally small time interval. You do this by taking the first derivative of the position vector with respect to time, giving you the instantaneous velocity vector ($\vec{v}$).

$$\vec{v} = \lim_{\Delta t \to 0} \frac{\Delta \vec{r}}{\Delta t} = \frac{d\vec{r}}{dt} = \vec{r}\,'(t)$$

Looking at the instantaneous velocity function in terms of x-, y-, and z-components, it can be written as:

$$\vec{v} = \frac{d\vec{r}}{dt} = <\frac{dx}{dt}, \frac{dy}{dt}, \frac{dz}{dt}> = \frac{dx}{dt}\hat{i} + \frac{dy}{dt}\hat{j} + \frac{dz}{dt}\hat{k}$$

You can gain a better understanding by further analyzing the position vs. time graph from a previous problem, shown below at left. Notice in the original graph on the left, as you look in the region between 1 and 6 seconds, the rate at which the object's location is changing is not constant. This represents an object moving with a velocity that is not constant, though the average velocity of the particle is 0.5 m/s. Looking to the graph on the right, however, you observe the line tangent to the curve having various slopes between 1 and 6 seconds, consistent with the object having varying velocities between 1 and 6 seconds.

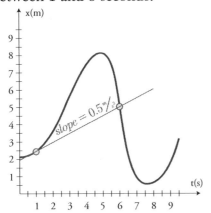
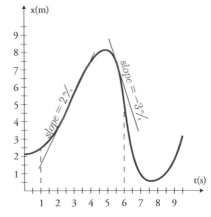

This also works in reverse. If you know the velocity of an object as a function of time, you can determine its displacement during that time interval. From an equation standpoint, the integral of the velocity gives you displacement. Written in one dimension (which you can, of course, extend to multiple dimensions as needed):

$$\int v(t)\,dt = x(t) + C$$

Graphically, the area under the velocity-time graph (below left) gives you the displacement function. If you wanted the displacement of the object from 1 to 6 seconds, you calculate the area under the graph from t=1 to t=6 seconds, which is equivalent to integrating the velocity function from t=1 to t=6 seconds, shown below right.

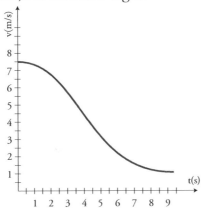
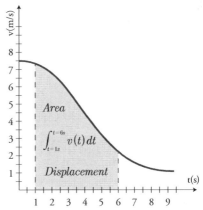

Following the same pattern, an object's instantaneous acceleration, the rate of change of an object's velocity at a specific instant in time, is found by taking the average acceleration over an infinitesimally small time interval, which you can do easily by taking the first derivative of the velocity vector with respect to time, or the second derivative of the position vector with respect to time. This corresponds to taking the slope of the line tangent to the velocity-time graph.

$$\vec{a} = \lim_{\Delta t \to 0} \frac{\Delta \vec{v}}{\Delta t} = \frac{d\vec{v}}{dt} = \vec{v}\,'(t)$$

$$= \frac{d^2\vec{r}}{dt^2} = \vec{r}\,''(t)$$

Of course, you can also take the integral of acceleration, corresponding to the area under the acceleration-time graph, to obtain the velocity function. Written in one dimension:

$$\int a(t)\,dt = v(t) + C$$

Putting this altogether, you can move between position, velocity, and acceleration using calculus and graphing skills.

   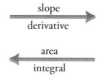

slope
derivative

area
integral

slope
derivative

area
integral

$\vec{r}(t)$ $\vec{v}(t)$ $\vec{a}(t)$

3.7 Q: The position of a particle as a function of time is: $x(t) = 2 - 4t + 2t^2 - 3t^3$. Find the velocity and acceleration of the particle as a function of time.

3.7 A: $v(t) = \dfrac{dx}{dt} = -9t^2 + 4t - 4$

$a(t) = \dfrac{dv}{dt} = \dfrac{d^2x}{dt^2} = -18t + 4$

3.8 Q: An object moving in a straight line has a velocity v in meters per second that varies with time t according to the following function: $v(t) = 3 + 2t^2$.
A) Determine the acceleration of the object at t=1 second.
B) Determine the displacement of the object between t=0 and t=5 seconds.

3.8 A: A) $a = \dfrac{dv}{dt} = 4t \xrightarrow{t=1s} a = 4(1) = 4\,{}^m\!/_{s^2}$

B) $\Delta x = \displaystyle\int_{t=0}^{5s} v(t)\,dt = \int_{0}^{5s} (3 + 2t^2)\,dt = \left[\dfrac{2t^3}{3} + 3t\right]\Big|_{0}^{5s} \rightarrow$

$\Delta x = \left[\dfrac{2(5)^3}{3} + 3(5)\right] - \left[\dfrac{2(0)^3}{3} + 3(0)\right] = 98m$

3.9 Q: The velocity-time curve for a tortoise and hare, traveling in a straight line, is shown below.

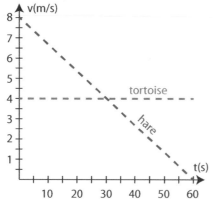

tortoise

hare

A) What happens at time t=30s?
B) How do you know the two have traveled the same distance at time t=60s?
C) What is the acceleration of the hare at t=40s?

3.9 A: A) At t=30s, the tortoise and the hare have the same speed.

B) At t=60s, the area under each curve is the same, so the distance traveled is the same.

C) The acceleration is the slope at t=40s: $a = slope = \dfrac{rise}{run} = \dfrac{0 - 8\,^m\!/_s}{60s} = -0.13\,^m\!/_{s^2}$

3.10 Q: Which of the following pairs of graphs best shows the distance traveled versus time and speed versus time for a race car accelerating down a hill from rest?

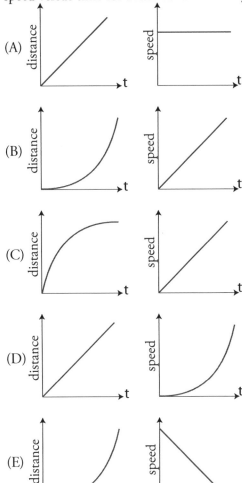

3.10 A: (B)

3.11 Q: An object's position is given by $x(t) = 8t^3 - 4t^2 + 3\sin(\pi t)$. Determine the object's average acceleration from time t=0 to t=8 seconds.

3.11 A: $v(t) = \dfrac{dx}{dt} = 24t^2 - 8t + 3\pi \cos(\pi t)$

$v(0) = 3\pi\,^m\!/_s$

$v(8) = 1481.4\,^m\!/_s$

$a_{avg} = \dfrac{\Delta v}{t} = \dfrac{v(8) - v(0)}{8s} = \dfrac{1481.4\,^m\!/_s - 3\pi\,^m\!/_s}{8s} = 184\,^m\!/_{s^2}$

3.12 Q: The acceleration of a particle moving in one dimension is given by the function $a(t) = 3t + 6$. Determine the position of the particle as a function of time. At time t=0, the particle is at position x=3m, and at time t=2s, the particle is at position x=25m.

3.12 A: $v(t) = \int a(t)\,dt = \int (3t+6)\,dt = \dfrac{3t^2}{2} + 6t + C$

$x(t) = \int v(t)\,dt = \int \left(\dfrac{3t^2}{2} + 6t + C\right)dt = \dfrac{t^3}{2} + 3t^2 + Ct + D$

Now use the given boundary conditions to determine the unknown constants C and D.

$x(t) = \dfrac{t^3}{2} + 3t^2 + Ct + D \xrightarrow{\; x(0)=3m \;} 3m = \dfrac{0^3}{2} + 3(0)^2 + C(0) + D \longrightarrow D = 3m$

$x(t) = \dfrac{t^3}{2} + 3t^2 + Ct + 3 \xrightarrow{\; x(2)=25m \;} 25m = 19m + 2C \rightarrow C = 3m$

$x(t) = \dfrac{t^3}{2} + 3t^2 + 3t + 3$

3.13 Q: Several motion graphs are shown below left. Using your knowledge of kinematics, sketch a possible corresponding motion graph on the right.

A)
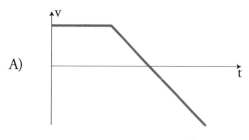

B)
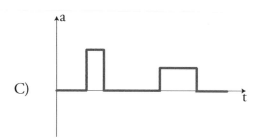

C)

Note: There are multiple correct solutions to parts B and C.

Chapter 3: Kinematics

3.13 A:

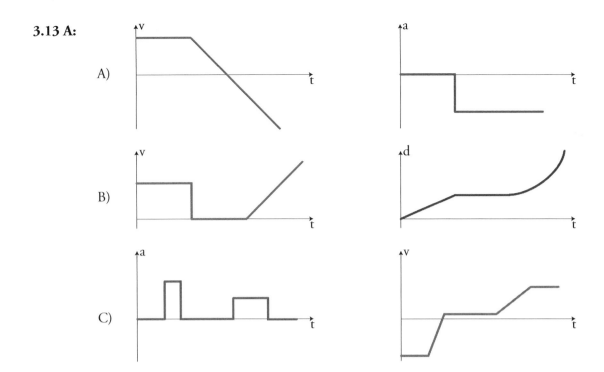

A)

B)

C)

Particle Diagrams

Graphs and diagrams are terrific tools for understanding physics, and they are especially helpful for studying motion, a phenomenon that we are used to perceiving visually. Particle diagrams, sometimes referred to as ticker-tape diagrams or dot diagrams, show the position or displacement of an object at evenly spaced time intervals.

Think of a particle diagram like an oil drip pattern. If a car has a steady oil drip, where one drop of oil falls to the ground every second, the pattern of the oil droplets on the ground could represent the motion of the car with respect to time. By examining the oil drop pattern, a bystander could draw conclusions about the displacement, velocity, and acceleration of the car, even if he wasn't able to watch the car drive by! The oil drop pattern is known as a particle diagram, ticker-tape diagram, or motion map.

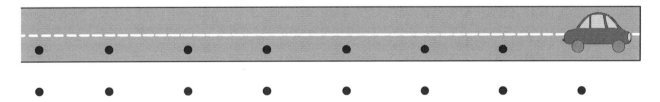

From the particle diagram above you can see that the car was moving either to the right or the left, and since the drops are evenly spaced, you can say with certainty that the car was moving at a constant velocity, and since velocity isn't changing, acceleration must be 0. So what would the particle diagram look like if the car was accelerating to the right? Take a look below and see!

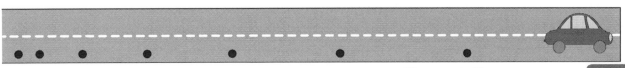

The oil drops start close together on the left, and get further and further apart as the object moves toward the right. Of course, this pattern could also have been produced by a car moving from right to left, beginning with a high velocity at the right and slowing down as it moves toward the left. Because the velocity vector (pointing to the left) and the acceleration vector (pointing to the right) are in opposite directions, the object slows down. This is a case where, if you called to the right the positive direction, the car would have a negative velocity, a positive acceleration, and it would be slowing down. Check out the resulting particle diagram below!

Can you think of a case in which the car could have a negative velocity and a negative acceleration, yet be speeding up? Draw a picture of the situation!

3.14 Q: The diagram below shows a car beginning from rest and accelerating uniformly down the road.

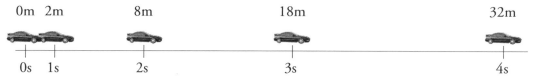

What is the average speed of the car between 2 and 4 seconds?

3.14 A: $\bar{v} = \dfrac{\Delta x}{t} = \dfrac{24m}{2s} = 12\,^m/_s$

Motion in Multiple Dimensions

By now you may have noticed that all the example problems were one dimensional. Knowing how to transition between position, velocity, and acceleration in one dimension allows you to transition among the same quantities in multiple dimensions. Using standard Cartesian (x, y, z) coordinates, you can write the equations for velocity and acceleration as:

$$\vec{v} = \frac{d\vec{r}}{dt} = \frac{dx}{dt}\hat{i} + \frac{dy}{dt}\hat{j} + \frac{dz}{dt}\hat{k} = <\frac{dx}{dt}, \frac{dy}{dt}, \frac{dz}{dt}>$$

$$\vec{a} = \frac{d\vec{v}}{dt} = \frac{d^2x}{dt^2}\hat{i} + \frac{d^2y}{dt^2}\hat{j} + \frac{d^2z}{dt^2}\hat{k} = <\frac{d^2x}{dt^2}, \frac{d^2y}{dt^2}, \frac{d^2z}{dt^2}>$$

3.15 Q: A particle's position is $\vec{r} = < 3t^3 - 6, t^2 + 2t, t - 1 > = (3t^3 - 6)\hat{i} + (t^2 + 2t)\hat{j} + (t - 1)\hat{k}$.
A) Determine the particle's velocity vector.
B) Determine the speed of the particle at time t=2 seconds.
C) Determine the particle's acceleration vector.
D) Determine the magnitude of the particle's acceleration at t=1 second.

3.15 A: A) $\vec{v} = \dfrac{d\vec{r}}{dt} = 9t^2\hat{i} + (2t+2)\hat{j} + \hat{k} = \;<9t^2, 2t+2, 1>$

B) $\vec{v}(t=2s) = \;<36,6,1> \to |\vec{v}| = \sqrt{36^2 + 6^2 + 1^2} = 36.5\,{}^{m}\!/\!_{s}$

C) $\vec{a} = \dfrac{d\vec{v}}{dt} = 18t\hat{i} + 2\hat{j} = \;<18t,2,0>$

D) $\vec{a}(t=1s) = \;<18,2,0> \to |\vec{a}| = \sqrt{18^2 + 2^2} = 18.1\,{}^{m}\!/\!_{s^2}$

3.16 Q: A superhero firefly's velocity is given by $\vec{v}(t) = 3t^2\hat{i} - 4\hat{j} + 2\sin(t)\hat{k}$.
A) Determine the firefly's displacement between t=0 and t=3 seconds.
B) Determine the acceleration of the firefly at t=2 seconds.

3.16 A: A) The displacement is the integral of the velocity from t=0 to t=3 seconds.
$$\Delta\vec{r} = \int_{t=0}^{3s} \vec{v}(t)\,dt = \int_{t=0}^{3s}(3t^2\hat{i} - 4\hat{j} + 2\sin(t)\hat{k})\,dt \Rightarrow$$

You can break up this integral recognizing the integral of a sum is the sum of the integrals.
$$\Delta\vec{r} = \int_{t=0}^{3s} 3t^2\hat{i}\,dt - \int_{t=0}^{3s} 4\hat{j}\,dt + \int_{t=0}^{3s} 2\sin(t)\hat{k}\,dt \Rightarrow$$

Those troublesome-looking unit vectors are quite easy to deal with once you recognize they are constants and can be pulled out of the integral sign.
$$\Delta\vec{r} = \hat{i}\int_{t=0}^{3s} 3t^2\,dt - \hat{j}\int_{t=0}^{3s} 4\,dt + \hat{k}\int_{t=0}^{3s} 2\sin(t)\,dt \Rightarrow$$

Now you can integrate the individual terms.
$$\Delta\vec{r} = t^3\hat{i} - 4t\hat{j} - 2\cos(t)\hat{k}\,\big|_{t=0}^{t=3s} = [(3^3)\hat{i} - 4(3)\hat{j} - 2\cos(3)\hat{k}] - [-2\cos(0)\hat{k}] \Rightarrow$$

$$\Delta\vec{r} = 27\hat{i} - 12\hat{j} + (2 - 2\cos(3))\hat{k} = \;<27, -12, 2-2\cos(3)>$$

B) $\vec{a} = \dfrac{d\vec{v}}{dt} = \;<6t, 0, 2\cos(t)> \xrightarrow{t=2s} \vec{a}(t=2s) = \;<12, 0, 2\cos(2)>$

Uniformly Accelerated Motion

The special case of constant acceleration is known as uniformly accelerated motion, or UAM. This condition is very common in everyday life, and many problems can be modeled quite accurately as exhibiting a constant acceleration, or breaking a problem into several smaller problems, each exhibiting a constant acceleration. In treating situations of uniformly accelerated motion, you can utilize a set of problem-solving equations in your physics toolbox, known as the kinematic equations. These equations can help you solve for key variables describing the motion of an object when you have a constant acceleration. Once you know the value of any three variables, you can use the kinematic equations to solve for the other two!

Variable	Meaning
v_0	initial velocity
v	final velocity
Δx	displacement
a	acceleration
t	time

$$v = v_0 + at$$
$$x = x_0 + v_0 t + \tfrac{1}{2} at^2$$
$$v^2 = v_0^2 + 2a(x - x_0)$$

Hint: Though the kinematic equations are given in terms of x, they can be applied to any of the coordinate axes.

In using these equations to solve motion problems, it's important to take care in setting up your analysis before diving into a solution. Key steps to solving kinematics problems include:

1. Labeling your analysis for horizontal (x-axis) or vertical (y-axis) motion.
2. Choosing and indicating a positive direction (typically the direction of initial motion).
3. Creating a motion analysis table (v_0, v, Δx, a, t). Note that Δx is a change in position, or displacement, and can be re-written as x-x_0.
4. Using what you know about the problem to fill in your "givens" in the table.
5. Once you know three items in the table, using kinematic equations to solve for any unknowns.
6. Verifying that your solution makes sense.

Take a look at a sample problem to see how this strategy can be employed.

3.17 Q: A race car starting from rest accelerates uniformly at a rate of 4.90 m/s². What is the car's speed after it has traveled 200 meters?
(A) 1960 m/s
(B) 62.6 m/s
(C) 44.3 m/s
(D) 31.3 m/s
(E) 40.8 m/s

3.17 A: Step 1: Horizontal Problem
Step 2: Positive direction is direction car starts moving.
Step 3 & 4:

Variable	Value
v_0	0 m/s
v	FIND
Δx	200 m
a_x	4.90 m/s²
t	?

Step 5: Choose a kinematic equation that includes the given information and the information sought, and solve for the unknown showing the initial formula, substitution with units, and answer with units.

$$v^2 = v_0^2 + 2a(x - x_0) \xrightarrow{v_0 = 0} v^2 = 2a(x - x_0) \xrightarrow[x - x_0 = 200m]{a = 4.90 \frac{m}{s^2}}$$
$$v^2 = 2(4.90 \tfrac{m}{s^2})(200m) = 1960 \tfrac{m^2}{s^2} \rightarrow v = \sqrt{1960 \tfrac{m^2}{s^2}} = 44.3 \tfrac{m}{s}$$

Step 6: (C) 44.3 m/s is one of the given answer choices, and is a reasonable speed for a race car (44.3 m/s is approximately 99 miles per hour).

This strategy also works for vertical motion problems.

3.18 Q: An astronaut standing on a platform on the Moon drops a hammer. If the hammer falls 6.0 meters vertically in 2.7 seconds, what is its acceleration?
(A) 0.82 m/s²
(B) 1.6 m/s²
(C) 2.2 m/s²
(D) 4.4 m/s²
(E) 9.8 m/s²

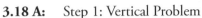

3.18 A: Step 1: Vertical Problem
Step 2: Positive direction is down (direction hammer starts moving).
Step 3 & 4: Note that a dropped object has an initial vertical velocity of 0 m/s.

Variable	Value
v_0	0 m/s
v	?
Δy	6 m
a	FIND
t	2.7 s

Step 5: Choose a kinematic equation that includes the given information and the information sought, and solve for the unknown showing the initial formula, substitution with units, and answer with units.

$$\Delta y = v_0 t + \tfrac{1}{2}at^2 \xrightarrow{v_0=0} \Delta y = \tfrac{1}{2}at^2 \rightarrow a = \frac{2\Delta y}{t} \xrightarrow[t=2.7s]{\Delta y=6m} a = \frac{2(6m)}{(2.7s)^2} = 1.6\,{}^m/_{s^2}$$

Step 6: (B) 1.6 m/s² is one of the given answer choices, and is less than the acceleration due to gravity on the surface of the Earth (9.8 m/s²). This answer can be verified further by searching on the Internet to confirm that the acceleration due to gravity on the surface of the moon is indeed 1.6 m/s².

© Fernando Gregory | Dreamstime.com

3.19 Q: An astronaut drops a hammer from 2.0 meters above the surface of the Moon. If the acceleration due to gravity on the Moon is 1.62 m/s², how long will it take for the hammer to fall to the Moon's surface?
(A) 0.62 s
(B) 1.2 s
(C) 1.6 s
(D) 2.5 s
(E) it will float forever

3.19 A: (C) $\Delta y = v_0 t + \tfrac{1}{2}at^2 \xrightarrow{v_0=0} \Delta y = \tfrac{1}{2}at^2 \rightarrow$
$$t = \sqrt{\frac{2\Delta y}{a}} = \sqrt{\frac{2(2m)}{1.62\,{}^m/_{s^2}}} = 1.57s$$

In some cases, you may not be able to solve directly for the "find" quantity. In these cases, you can solve for the other unknown variable first, then choose an equation to give you your final answer.

3.20 Q: Bill, driving on a straight road at 15.0 meters per second, accelerates uniformly to a speed of 21.0 meters per second in 12.0 seconds. The total distance traveled by Bill in this 12.0-second time interval is
(A) 36.0 m
(B) 180 m
(C) 216 m
(D) 252 m
(E) 315 m

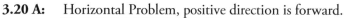

3.20 A: Horizontal Problem, positive direction is forward.

Variable	Value
v_0	15 m/s
v	21 m/s
Δx	FIND
a	?
t	12 s

Can't find Δx directly, find a first.

$$v = v_0 + at \rightarrow a = \frac{v - v_0}{t} \rightarrow a = \frac{21\,m/s - 15\,m/s}{12s} = 0.5\,m/s^2$$

Now solve for Δx.

$$\Delta x = v_0 t + \tfrac{1}{2} at^2 = (15\,m/s)(12s) = \tfrac{1}{2}(0.5\,m/s^2)(12s)^2 = 216m$$

Check: (C) 216m is a given answer, and is reasonable, as this is greater than the 180m Bill would have traveled if remaining at a constant speed of 15 m/s for the 12 second time interval.

Catch-Up and Interpretation Problems

Now that you have the basics down, you can begin to tackle more advanced problems, such as problems involving two or more objects moving simultaneously, or problems that eschew numeric givens for symbolic givens. The key to solving these problems is taking your time in interpreting the given information and translating into mathematical relationships. Some examples are given below.

3.21 Q: Rush, the crime-fighting superhero, can run at a maximum speed of 30 m/s, while Evil Eddie, the criminal mastermind, can run 5 m/s. If Evil Eddie is 500 m ahead of Rush and both are running at their maximum speeds:
A) How much time does Evil Eddie have to devise an escape plan?
B) How far must Rush run to capture Evil Eddie?

3.21 A: Start by setting up the problem. Let x_R represent Rush's distance from the origin, and x_{EE} represent Evil Eddie's distance from the origin. You can then model their positions as functions of time:
$$x_R = 30t$$
$$x_{EE} = 500 + 5t$$

A) Evil Eddie runs out of time when they have the same position, so set x_R equal to x_{EE} in order to solve for the time when they meet.
$$x_R = x_{EE} \rightarrow 30t = 500 + 5t \rightarrow t = 20s$$

B) If Rush catches Evil Eddie in 20s, you can find the distance Rush must run by multiplying his maximum velocity by the time he must run.
$$x_R = 30t = \left(30\,{}^m\!/_s\right)(20s) = 600m$$

3.22 Q: A swimmer swims three-fifths the width of a river at one velocity (v), then swims the remainder of the river at half her initial velocity (½v). What was her average speed swimming across the river?
(A) 0.71v
(B) 0.75v
(C) 0.80v
(D) 0.88v

3.22 A: (A) Begin by finding the time for each leg of the journey, then determine the average speed for the entire trip across the width of the river (r).
$$t_A = \frac{\Delta s}{v} = \frac{0.6r}{v}$$

$$t_B = \frac{\Delta s}{v} = \frac{0.4r}{0.5v} = \frac{0.8r}{v}$$

$$\bar{v} = \frac{\Delta s}{t} = \frac{r}{t_A + t_B} = \frac{r}{0.6\frac{r}{v} + 0.8\frac{r}{v}} = 0.71v$$

*3.23 Q: Car A is traveling 40 m/s when it crosses a dashed line in the road at time t=0. As soon as Car A crosses the dashed line, the driver hits the brakes, causing Car A to decelerate at 5 m/s². Car B begins its journey from rest at the dashed line. At time t=0, Car B accelerates in the same direction that Car A is traveling, with an acceleration given by the formula $a_B = 10 - t$.
A) How long does it take Car B to catch up to Car A?
B) How far from the dashed line do the cars meet?

*3.23 A: A) Since you're given the acceleration of Car B as a function of time, you can begin the problem by solving for Car B's velocity function. Note the use of boundary conditions to determine the constant of integration, C.
$$v_B(t) = \int a_B(t)\,dt = \int (10 - t)\,dt = 10t - \frac{t^2}{2} + C \xrightarrow[C=0]{v(0)=0} v_B(t) = 10t - \frac{t^2}{2}$$

Next you can find Car B's position function by integration the velocity function, again using your boundary conditions to determine the constant of integration.
$$x_B(t) = \int \left(10t - \frac{t^2}{2}\right)dt = \frac{10t^2}{2} - \frac{1}{2}\frac{t^3}{3} + C \xrightarrow[C=0]{x(0)=0} x_B(t) = 5t^2 - \frac{t^3}{6}$$

You can also write a position function for Car A using your kinematic equations, as Car A exhibits a constant acceleration.

$$x_A(t) = v_{A_0}(t) + \tfrac{1}{2}a_A t^2 = 40t + \tfrac{1}{2}(-5)t^2 = 40t - 2.5t^2$$

Car B catches up to Car A when they have the same position, so set the positions of the two cars as equal to develop an equation in terms of just the elapsed time, t.

$$x_A(t) = x_B(t) \rightarrow 40t - 2.5t^2 = 5t^2 - \frac{t^3}{6} \rightarrow \frac{t^3}{6} - 7.5t^2 + 40t = 0$$

From here, you can factor out t, recognizing that, of course, time t=0 is one solution to the equation (which we already should have known, as of course the cars are at the same position at time t=0). However, you want to know at what time after t=0 the cars are at the same position, which can be determined by solving the resulting quadratic equation.

$$\frac{t^3}{6} - 7.5t^2 + 40t = 0 \rightarrow t\left(\frac{t^2}{6} - 7.5t + 40\right) = 0 \xrightarrow{t=0 \ is \ solution} \frac{t^2}{6} - 7.5t + 40 = 0$$

There are a variety of strategies to solving quadratic equations, the most popular, of course, being the quadratic equation. As you are preparing for the AP Physics C exam, however, where time is critical and calculators are allowed, I would strongly recommend learning how to solve this equation using the exact model of calculator you're planning on using for the exam. Regardless of method, you should find answers of t=6.18s and t=38.8s.

Common sense should tell you the first time, t=6.18s, is when Car B first catches up to Car A (the second time describes the point at which the cars would be at the same position at the same time again, in this case, when they are both traveling backwards). The wording of the problem indicates that car A will stop after its velocity reaches 0, however (at time t=8 seconds), so any solutions with a time greater than 8 seconds are not appropriate for this problem, and therefore t=6.18s is the correct answer.

B) You can find the distance from the dashed line to the catch-up point using either Car A or Car B's position function with a time of t=6.18s:

$$x_A(t) = 40t - 2.5t^2 \xrightarrow{t=6.18s} x_A(t) = 40(6.18) - 2.5(6.18^2) = 152m$$

Free Fall

Examination of free-falling bodies dates back to the days of Aristotle. At that time Aristotle believed that more massive objects would fall faster than less massive objects. He believed this in large part due to the fact that when examining a rock and a feather falling from the same height it is clear that the rock hits the ground first. Upon further examination it is clear that Aristotle was incorrect in his hypothesis.

As proof, take a basketball and a piece of paper. Drop them simultaneously from the same height... do they land at the same time? Probably not. Now take that piece of paper and crumple it up into a tight ball and repeat the experiment. Now what do you see happen? You should see that both the ball and the paper land at the same time. Therefore you can conclude that Aristotle's predictions did not account for the effect of air resistance.

In the 17th century, Galileo Galilei began a re-examination of the motion of falling bodies. Galileo, recognizing that air resistance affects the motion of a falling body, executed his famous thought experiment in which he continuously asked what would happen if the effect of air resistance were removed. Commander David Scott of Apollo 15 performed this experiment while on the moon. He simultaneously dropped a hammer and a feather, and observed that they reached the ground at the same time.

Since Galileo's experiments, scientists have come to a better understanding of how the gravitational pull of the Earth accelerates free-falling bodies. Through experimentation it has been determined that the local **gravitational field strength (g)** on the surface of the Earth is 9.8 N/kg, which further indicates that all objects in free fall (neglecting air resistance) experience an equivalent acceleration of 9.8 m/s² toward the center of the Earth.

> **Hint**: If you move off the surface of the Earth, the local gravitational field strength, and therefore the acceleration due to gravity, changes. For the purposes of the AP Physics C exam, it is acceptable to estimate g as 10 m/s² to simplify calculations.

We will analyze the motion of two free fall conditions: objects dropped from rest, and objects launched vertically upward.

Objects Falling From Rest

Objects starting from rest have an initial velocity of zero, giving you your first kinematic quantity needed for problem solving. Beyond that, if you call the direction of initial motion (down) positive, the object will have a positive acceleration and speed up as it falls.

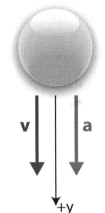

An important first step in analyzing objects in free fall is deciding which direction along the y-axis you are going to call positive and which direction will therefore be negative. Although you can set your positive direction any way you want and get the correct answer, following the hints below can simplify your work to reach the correct answer consistently.

1. Identify the direction of the object's initial motion and assign that as the positive direction. In the case of a dropped object, the positive y-direction will point toward the bottom of the paper.
2. With the axis identified you can now identify and write down your given kinematic information. Don't forget that a dropped object has an initial velocity of zero.
3. Notice the direction the vector arrows are drawn — if the velocity and acceleration point in the same direction, the object speeds up. If they point in opposite directions, the object slows down.

3.24 Q: What is the speed of a 2.5-kilogram mass after it has fallen freely from rest through a distance of 15 meters? Neglect air resistance.
(A) 4.8 m/s
(B) 17 m/s
(C) 30 m/s
(D) 43 m/s
(E) 50 m/s

3.24 A: Vertical Problem: Declare down as the positive direction. This means that the acceleration, which is also down, is a positive quantity.

Variable	Value
v_0	0 m/s
v	FIND
Δy	15 m
a	10 m/s²
t	?

$$v^2 = v_0^2 + 2a\Delta y = 2\left(10\tfrac{m}{s^2}\right)(15m) = 300\tfrac{m^2}{s^2} \rightarrow v = \sqrt{300\tfrac{m^2}{s^2}} = 17.3\tfrac{m}{s}$$

Correct answer is (B) 17 m/s.

3.25 Q: How far will a brick starting from rest fall freely in 3.0 seconds? Neglect air resistance.
(A) 15 m
(B) 29 m
(C) 45 m
(D) 66 m
(E) 90 m

3.25 A: (C) 45m

Variable	Value
v_0	0 m/s
v	?
Δy	FIND
a	10 m/s²
t	3 s

$$\Delta y = v_0^t + \tfrac{1}{2}at^2 \rightarrow \Delta y = \tfrac{1}{2}\left(10\tfrac{m}{s^2}\right)(3s)^2 = 45m$$

Objects Launched Upward

Examining the motion of an object launched vertically upward is done in much the same way you examined the motion of an object falling from rest. The major difference is that you have to look at two segments of its motion instead of one: both up *and* down.

Before you get into establishing a frame of reference and working through the quantitative analysis, you must build a solid conceptual understanding of what is happening while the ball is in the air. Consider the ball being thrown vertically into the air as shown in the diagram.

In order for the ball to move upwards, its initial velocity must be greater than zero. As the ball rises, its velocity decreases until it reaches its maximum height, where it stops and then begins to fall. As the ball falls, its speed increases. In other words, the ball is accelerating the entire time it is in the air: both on the way up, at the instant it stops at its highest point, and on the way down.

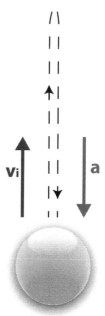

The cause of the ball's acceleration is gravity. The entire time the ball is in the air, its acceleration is 9.8 m/s² down provided this occurs on the surface of the Earth. Note that the acceleration can be either 9.8 m/s² or -9.8 m/s². The sign of the acceleration depends on the direction you declared as positive, but in all cases the direction of the acceleration due to gravity is down, toward the center of the Earth.

You have already established the ball's acceleration for the entire time it is in the air is 9.8 m/s² down. This acceleration causes the ball's velocity to decrease at a constant rate until it reaches maximum altitude, at which point it turns around and starts to fall. In order to turn around, the ball's velocity must pass through zero. Therefore, at maximum altitude the velocity of the ball must be zero.

Because gravity provides the same acceleration to the ball on the way up (slowing it down) as on the way down (speeding it up), the time to reach maximum altitude is the same as the time to return to its launch position. In similar fashion, the initial speed of the ball on the way up will equal the speed of the ball at the instant it reaches the point from which it was launched on the way down. Put another way, the time to go up is equal to the time to go down, and the initial speed up is equal to the final speed down (assuming the object begins and ends at the same height above ground).

Now that a conceptual understanding of the ball's motion has been established, you can work toward a quantitative solution. Following the rule of thumb established previously, you can start by assigning the direction the ball begins to move as positive. Remember that assigning positive and negative directions are completely arbitrary. You have the freedom to assign them how you see fit. Once you assign them, however, don't change them.

Once this positive reference direction has been established, all other velocities and displacements are assigned accordingly. For example, if up is the positive direction, the acceleration due to gravity will be negative, because the acceleration due to gravity points down, toward the center of the Earth. At its highest point, the ball will have a positive displacement, and will have a zero displacement when it returns to its starting point. If the ball isn't caught, but continues toward the Earth past its starting point, it will have a negative displacement.

A "trick of the trade" to solving free fall problems involves symmetry. As long as you are neglecting air resistance, the time an object takes to reach its highest point is equal to the time it takes to return to the same vertical position. The speed with which the projectile begins its journey upward is equal to the speed of the projectile when it returns to the same height (although, of course, its velocity is in the opposite direction). If you want to simplify the problem, vertically, at its highest point, the vertical velocity is 0. This added information can assist you in filling out your vertical motion table. If you cut the object's motion in half, you can simplify your problem solving – but don't forget that if you want the total time in the air, you must double the time it takes for the object to rise to its highest point.

3.26 Q: A basketball player jumped straight up to grab a rebound. If she was in the air for 0.80 seconds, how high did she jump? Neglect air resistance.

3.26 A: Define up as the positive y-direction. Note that if basketball player is in the air for 0.80 seconds, she reaches her maximum height at a time of 0.40 seconds, at which point her velocity is zero.

Variable	Value
v_0	?
v	0 m/s
Δy	FIND
a	-10 m/s^2
t	0.40 s

Can't solve for Δy directly with given information, so find v_0 first.
$$v = v_0 + at \rightarrow v_0 = v - at = 0 - (-10\,\tfrac{m}{s^2})(0.40s) = 4\,\tfrac{m}{s}$$
Now with v_0 known, solve for Δy.
$$\Delta y = v_0 t + \tfrac{1}{2}at^2 = (4\,\tfrac{m}{s})(0.40s) + \tfrac{1}{2}(-10\,\tfrac{m}{s^2})(0.40s)^2 = 0.8m$$
This is a reasonable height for a basketball player to jump.

If, instead, you analyze the motion on the way down, you find that v_0=0, a=10 m/s^2, and t=0.4 s. Solving for displacement:
$$\Delta y = v_0 t + \tfrac{1}{2}at^2 = (0)(0.40s) + \tfrac{1}{2}(10\,\tfrac{m}{s^2})(0.40s)^2 = 0.8m$$
You find the same answer, but with significantly less work!

3.27 Q: Three model rockets of varying masses are launched vertically upward from the ground with varying initial velocities. From highest to lowest, rank the maximum height reached by each rocket. Neglect air resistance.

m = 800 g
v = 40 m/s

m = 1 kg
v = 20 m/s

m = 1500 g
v = 30 m/s

(A) 1, 2, 3
(B) 3, 1, 2
(C) 3, 2, 1
(D) 1, 3, 2
(E) 2, 1, 3

3.27 A: (D) 1, 3, 2. The acceleration due to gravity is independent of mass, therefore the object with the largest initial vertical velocity will reach the highest maximum height.

Projectile Motion

Projectile motion problems, or problems of an object launched in both the x- and y- directions and neglecting air resistance, can be analyzed using the physics you already know. The key to solving these types of problems is realizing that the horizontal component of the object's motion is independent of the vertical component of the object's motion. Since you already know how to solve horizontal and vertical kinematics problems, all you have to do is put the two results together!

Start these problems by making separate motion tables for vertical and horizontal motion. Vertically, the setup is the same for projectile motion as it is for an object in free fall. Horizontally, gravity only pulls an object down, it never pulls or pushes an object horizontally; therefore, the horizontal acceleration of the projectile is zero. If the acceleration horizontally is zero, velocity must be constant; therefore v_0 horizontally must equal v horizontally. Finally, to tie the problem together, realize that the time the projectile is in the air vertically must be equal to the time the projectile is in the air horizontally. This results in a parabolic path for the object, known as a **trajectory**.

When an object is launched or thrown completely horizontally, such as a rock thrown horizontally off a cliff, the initial velocity of the object is its initial horizontal velocity. Because horizontal velocity doesn't change, this velocity is also the object's final horizontal velocity, as well as its average horizontal velocity. Further, the initial vertical velocity of the projectile is zero. This means that you could hurl an object 1000 m/s horizontally off a cliff, and simultaneously drop an object off the cliff from the same height, and they will both reach the ground at the same time (even though the hurled object has traveled a greater distance).

3.28 Q: Fred throws a baseball 42 m/s horizontally from a height of two meters. How far will the ball travel before it reaches the ground?

3.28 A: To solve this problem, you must first find how long the ball will remain in the air. This is a vertical motion problem.

Vertical	Value
v_{0y}	0 m/s
v_y	?
Δy	2 m
a_y	10 m/s²
t	FIND

$$\Delta y = v_{0_y}t + \tfrac{1}{2}a_yt^2 \xrightarrow{v_{0_y}=0} \Delta y = \tfrac{1}{2}a_yt^2 \rightarrow$$

$$t = \sqrt{\frac{2\Delta y}{a_y}} = \sqrt{\frac{2\,(2m)}{10\,^m\!/_{s^2}}} = 0.63s$$

Now that you know the ball is in the air for 0.63 seconds, you can find how far it travels horizontally before reaching the ground. This is a horizontal motion problem, in which the acceleration is zero (nothing is causing the ball to accelerate horizontally). Because the ball doesn't accelerate, its initial horizontal velocity is also its final horizontal velocity, which is equal to its average horizontal velocity.

Horizontal	Value
v_{0x}	42 m/s
v_x	42 m/s
Δx	FIND
a_x	0 m/s²
t	0.63 s

$$\Delta x = v_{0_x}t + \tfrac{1}{2}a_xt^2 \xrightarrow{a_x=0} \Delta x = v_{0_x}t = (42\,^m\!/_s)(0.63s) = 26m$$

You can therefore conclude that the baseball travels 26 meters horizontally before reaching the ground.

3.29 Q: A 0.2-kilogram red ball is thrown horizontally at a speed of 4 meters per second from a height of 3 meters. A 0.4-kilogram green ball is thrown horizontally from the same height at a speed of 8 meters per second. Compared to the time it takes the red ball to reach the ground, the time it takes the green ball to reach the ground is
(A) one-half as great
(B) twice as great
(C) the same
(D) four times as great
(E) cannot be determined

3.29 A: (C) the same. Both objects are thrown horizontally from the same height. Because horizontal motion and vertical motion are independent, both objects have the same vertical motion (they both start with an initial vertical velocity of 0 m/s, have the same acceleration of 9.8 m/s² down, and both travel the same vertical distance). Therefore, the two objects reach the ground in the same amount of time.

Angled Projectiles

For objects launched at an angle, you have to do a little more work to determine the initial velocity in both the horizontal and vertical directions. For example, if a football is kicked with an initial velocity of 40 m/s at an angle of 30° above the horizontal, you need to break the initial velocity vector up into x- and y-components in the same manner as covered in the components of vectors math review section.

Then, use the components for your initial velocities in your horizontal and vertical tables. Finally, don't forget that symmetry of motion also applies to the parabola of projectile motion. For objects launched and landing at the same height, the launch angle is equal to the landing angle, and the launch speed is equal to the landing speed. If you want an object to travel the maximum possible horizontal distance (or range), launch it at an angle of 45°.

3.30 Q: Herman the Human Cannonball is launched from level ground at an angle of 30° above the horizontal with an initial velocity of 26 m/s. How far does Herman travel horizontally before reuniting with the ground?

3.30 A: The first step in solving this type of problem is to determine Herman's initial horizontal and vertical velocity. You do this by breaking up his initial velocity vector into vertical and horizontal components:
$v_{0_x} = v_0 \cos(\theta) = (26\,^m/_s) \cos(30°) = 22.5\,^m/_s$
$v_{0_y} = v_0 \sin(\theta) = (26\,^m/_s) \sin(30°) = 13\,^m/_s$

Next, analyze Herman's vertical motion to find out how long he is in the air. You can analyze his motion on the way up, find the time, and double that to find his total time in the air:

Vertical	Value
v_{0y}	13 m/s
v_y	0 m/s
Δy	?
a_y	-10 m/s²
t	FIND

$v_y = v_{0_y} + a_y t \rightarrow t_{up} = \dfrac{v_y - v_{0_y}}{a_y} = \dfrac{0 - 13\,^m/_s}{-10\,^m/_{s^2}} = 1.3s \rightarrow t_{total} = 2.6s$

Now that you know Herman was in the air 2.6s, you can find how far he moved horizontally, using his initial horizontal velocity of 22.5 m/s.

Horizontal	Value
v_{0x}	22.5 m/s
v_x	22.5 m/s
Δx	FIND
a_x	0
t	2.6 s

$$\Delta x = v_{0_x}t + \tfrac{1}{2}a_x t^2 \xrightarrow{a_x=0} \Delta x = v_{0_x}t = (22.5\,^m\!/_s)(2.6s) = 59m$$

Herman must have traveled 59 meters horizontally before returning to the Earth.

3.31 Q: A 30° incline sits on a 1.1-meter high table. A ball rolls off the incline with a velocity of 2.00 m/s. How far does the ball travel across the room before reaching the floor?

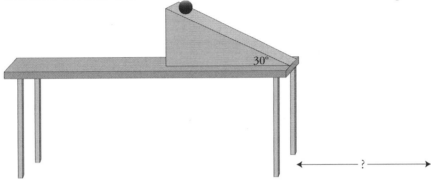

3.31 A: In order to determine how far the ball travels horizontally, you must first determine how long the ball is in the air. Begin by breaking up its initial velocity into horizontal and vertical components.
$$v_{0_x} = v_0 \cos(\theta) = (2.00\,^m\!/_s)\cos(30°) = 1.73\,^m\!/_s$$
$$v_{0_y} = v_0 \sin(\theta) = (2.00\,^m\!/_s)\sin(30°) = 1.00\,^m\!/_s$$

Determining how long the ball is in the air is a vertical motion problem. If you call the down direction positive, you can set up a vertical motion table as shown below.

Vertical	Value
v_{0y}	1.00 m/s
v_y	?
Δy	1.1 m
a_y	10 m/s²
t	FIND

To solve this problem, you could solve for time in the kinematic equation: $\Delta y = v_{0_y}t + \tfrac{1}{2}a_y t^2$. However, in doing so, you'll encounter a quadratic equation that will require you to utilize the quadratic formula. This is perfectly solvable, but you can save yourself some time and mathematical complexity if instead you solve for final velocity first, then solve for time.
$$v_y^2 = v_{0_y}^2 + 2a_y\Delta y = (1)^2 + 2(10)(1.1) = 23\,^{m^2}\!/_{s^2} \rightarrow v_y = \sqrt{23\,^{m^2}\!/_{s^2}} = 4.8\,^m\!/_s$$

Now, knowing the ball's final vertical velocity, you can solve for the time the ball is in the air: $v_y = v_{0_y} + a_y t \rightarrow t = \dfrac{v_y - v_{0_y}}{a_y} = \dfrac{4.8\,^m\!/_s - 1\,^m\!/_s}{10\,^m\!/_{s^2}} = 0.38s$

Next, knowing the time the ball is in the air, you can analyze the horizontal motion of the ball to calculate the horizontal distance traveled. Since you're neglecting air resistance, the horizontal acceleration of the ball is zero; therefore, the initial velocity is the same as the final velocity.

Horizontal	Value
v_{0x}	1.73 m/s
v_x	1.73 m/s
Δx	FIND
a_x	0
t	0.38 s

$$\Delta x = v_{0_x}t + \tfrac{1}{2}a_x t^2 \xrightarrow{a_x = 0} \Delta x = v_{0_x}t = (1.73\,^m/_s)(0.38s) = 0.66m$$

The ball travels 0.66m across the room horizontally from the edge of the table before striking the ground.

3.32 Q: A golf ball is hit at an angle of 45° above the horizontal. What is the acceleration of the golf ball at the highest point in its trajectory? [Neglect friction.]
(A) 9.8 m/s² upward
(B) 9.8 m/s² downward
(C) 6.9 m/s² horizontal
(D) 0 m/s²

3.32 A: (B) 9.8 m/s² downward

Path of a Projectile

There may be times when you want to sketch out the trajectory of a projectile. You can convert from the parametric equations describing the x- and y-coordinates as a function of time to an equation describing the y-coordinate as the function of the x-coordinate by solving the x-coordinate parametric equation for time. Then, substitute for time in the y-coordinate parametric equation.

3.33 Q: A daredevil rides his motorcycle off a ramp at a speed of 50 m/s. The ramp has an angle of 15 degrees above the horizontal, and ends 2 meters above the ground. Neglecting air resistance:
A) Determine the x- and y-coordinates of the daredevil as a function of time.
B) Determine the y-coordinate of the motorcycle as a function of the x-coordinate.
C) Sketch the motorcycle's trajectory.

3.33 A: A) Solve for the x- and y-components of the initial velocity:
$$v_{0_x} = (50\,^m/_s)\cos(15°) = 48.3\,^m/_s$$
$$v_{0_y} = (50\,^m/_s)\sin(15°) = 12.9\,^m/_s$$

Next solve for the y-displacement using the kinematic equations.
$$y = y_0 + v_{0_y}t + \tfrac{1}{2}a_y t^2 = 2 + 12.9t + \tfrac{1}{2}(-10)t^2$$

You can then solve for time using the x-displacement equation.

$$\Delta x = v_x t \rightarrow x = \left(48.3\,{}^m\!/\!_s\right)t \rightarrow t = \frac{x}{v_x} = \frac{x}{48.3}$$

B) Now you can substitute your expression for time as a function of the x-coordinate into the parametric equation of the y-coordinate.

$$y = 2 + 12.9t + \tfrac{1}{2}\left(-10\right)t^2 \xrightarrow{\;t=\frac{x}{48.3}\;} y = 2 + 12.9\left(\frac{x}{48.3}\right) + \tfrac{1}{2}\left(-10\right)\left(\frac{x}{48.3}\right)^2 \rightarrow$$

$$y = 2 + 0.267x - 0.00214x^2$$

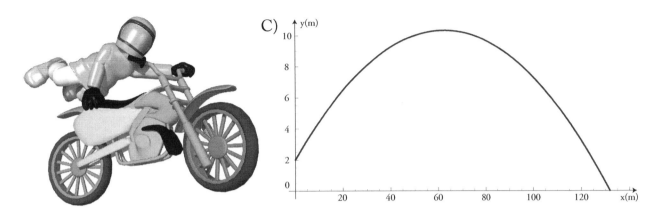

C)

3.34 Q: An aardvark launched from a cannon at an angle of 50° above the horizontal strikes a building 20 meters away at a point 10 meters above the point of projection.
A) Find the initial velocity of the aardvark.
B) Find the magnitude and direction of the aardvark's velocity vector just before it strikes the building.

3.34 A: A) Solve for the x- and y-components of the aardvark's initial velocity:

$$v_{0_x} = v_0 \cos\left(50°\right)$$
$$v_{0_y} = v_0 \sin\left(50°\right)$$

Next solve for time using the x-displacement equation.

$$\Delta x = v_{0_x} t \rightarrow t = \frac{\Delta x}{v_{0_x}} = \frac{20}{v_0 \cos\left(50°\right)}$$

Then you can solve for the y-displacement using the kinematic equations and your previous expression for time.

$$\Delta y = v_{0_y} t + \tfrac{1}{2} a_y t^2 \xrightarrow[\;v_{0_y}=v_0\sin(50°)\;]{\;t=\frac{20}{v_0\cos(50°)}\;} \Delta y = v_0 \sin\left(50°\right)\left(\frac{20}{v_0 \cos\left(50°\right)}\right) + \tfrac{1}{2} a_y t^2 \xrightarrow[\;a_y=-10\;]{\;\Delta y=10\;}$$

$$10 = 20\frac{\sin\left(50°\right)}{\cos\left(50°\right)} - 5\left(\frac{20}{v_0 \cos\left(50°\right)}\right)^2 \xrightarrow{\;\frac{\sin u}{\cos u}=\tan(u)\;} 2 = 4\tan\left(50°\right) - \frac{400}{v_0^2 \cos^2\left(50°\right)}$$

Now you can solve the resulting equation for v_0. Though it can be done by hand, this is another great opportunity to utilize the extended functionality of your calculator.

$$2 = 4\tan\left(50°\right) - \frac{400}{v_0^2 \cos^2\left(50°\right)} \rightarrow v_0 = 18.7\,{}^m\!/\!_s$$

B) Knowing the initial velocity, you can solve for the time at which the aardvark strikes the building.

$$t = \frac{\Delta x}{v_{0_x}} = \frac{20m}{\left(18.7\,{}^m\!/\!_s\right)\cos\left(50°\right)} = \frac{20m}{12\,{}^m\!/\!_s} = 1.66s$$

You found the horizontal velocity in the previous equation as 12 m/s. Find the vertical velocity at time t=1.66s using your kinematic equations.

$$v_y = v_{0_y} + at \xrightarrow[\substack{v_{0_y}=v_0\sin(50°) \\ a=-10}]{} v_y = v_0\sin(50°) - 10t \xrightarrow[\substack{v_0=18.7\,^m\!/_s \\ t=1.66s}]{}$$

$$v_y = (18.7\,^m\!/_s)\sin(50°) - 10(1.66s) = -2.3\,^m\!/_s$$

Now you can determine the magnitude and direction of the total velocity vector at time t=1.66s through vector composition.

$$|\vec{v}| = \sqrt{v_x^2 + v_y^2} = \sqrt{(12\,^m\!/_s)^2 + (-2.3\,^m\!/_s)^2} = 12.2\,^m\!/_s$$

$$\theta = \tan^{-1}\left(\frac{v_y}{v_x}\right) = \tan^{-1}\left(\frac{-2.3\,^m\!/_s}{12\,^m\!/_s}\right) = -11°$$

Therefore, the aardvark struck the building at a speed of 12.2 m/s and an angle of 11° below the horizontal.

Circular Motion

Now that you've talked about linear and projectile kinematics, you have the skills and background to analyze circular motion. Of course, this has obvious applications such as cars moving around a circular track, roller coasters moving in loops, and toy trains chugging around a circular track under the Christmas tree. Less obvious, however, is the application to turning objects. Any object that is turning can be thought of as moving through a portion of a circular path, even if it's just a small portion of that path.

With this realization, analysis of circular motion will allow you to explore a car speeding around a corner on an icy road, a running back cutting upfield to avoid a blitzing linebacker, and the orbits of planetary bodies. The key to understanding all of these phenomena starts with the study of uniform circular motion.

Radians and Degrees

Typically people discuss rotational motion in terms of degrees, where one entire rotation around a circle is equal to 360°. When dealing with rotational motion from a physics perspective, measuring rotational motion in units known as radians (rads) is much more efficient. A **radian** measures a distance around an arc equal to the length of the arc's radius.

Up to this point, you've described distances and displacements in terms of Δx, Δy, and Δz. In discussing **angular displacements**, you must transition to describing the distance traveled around an arc (known as the arc length) in terms of the variable s, while continuing to use the symbol θ (theta) to represent angles and angular displacement.

The distance completely around a circular path (360°), known as the **circumference**, C, can be found using $\Delta s = C = 2\pi r = 2\pi$ *radians*. Therefore, you can use this as a conversion factor to move back and forth between degrees and radians.

3.35 Q: Convert 90° to radians.

3.35 A: $90° \times \dfrac{2\pi \; radians}{360°} = \dfrac{\pi}{2} \; radians = 1.57 \; radians$

3.36 Q: Convert 6 radians to degrees.

3.36 A: $6 \; radians \times \dfrac{360°}{2\pi \; radians} = 344°$

Angles are also measured in terms of revolutions (complete trips around a circle). A complete single rotation is equal to 360°; therefore you can write the conversion factors for rotational distances and displacements as $360° = 2\pi \; radians = 1 \; revolution$.

3.37 Q: Convert 1.5 revolutions to both radians and degrees.

3.37 A: $1.5 \; revolutions \times \dfrac{2\pi \; radians}{1 \; revolution} = 3\pi \; radians$

$1.5 \; revolutions \times \dfrac{360°}{1 \; revolution} = 540°$

Uniform Circular Motion

The motion of an object in a circular path at constant speed is known as **uniform circular motion** (**UCM**). An object in UCM is constantly changing direction, and since velocity is a vector and has direction, you could say that an object undergoing UCM has a constantly changing velocity, even if its speed remains constant. And if the velocity of an object is changing, it must be accelerating. Therefore, an object undergoing UCM is constantly accelerating. This type of acceleration is known as **centripetal acceleration**.

3.38 Q: If a car is accelerating, is its speed increasing?

3.38 A: It depends. Its speed could be increasing, or it could be accelerating in a direction opposite its velocity (slowing down). Or, its speed could remain constant yet still be accelerating if it is traveling in uniform circular motion.

Just as importantly, you'll need to figure out the direction of the object's acceleration, since acceleration is a vector. To do this, draw an object moving counter-clockwise in a circular path, and show its velocity vector at two different points in time. Since acceleration is the rate of change of an object's velocity with respect to time, you can determine the direction of the object's acceleration by finding the direction of its change in velocity, Δv.

To find its change in velocity, Δv, recall that $\Delta \vec{v} = \vec{v} - \vec{v}_0$.

Therefore, you can find the difference of the vectors v and v_0 graphically, which can be re-written as $\Delta \vec{v} = \vec{v} + (-\vec{v}_0)$.

Recall that to add vectors graphically, you line them up tip-to-tail, then draw the resultant vector from the starting point (tail) of the first vector to the ending point (tip) of the last vector.

So, the acceleration vector must point in the direction shown above. If this vector is shown back on the original circle, lined up directly between the initial and final velocity vector, it's easy to see that the acceleration vector points toward the center of the circle.

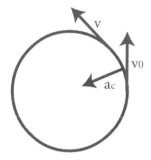

You can repeat this procedure from any point on the circle. No matter where you go, the acceleration vector always points toward the center of the circle. In fact, the word centripetal in centripetal acceleration means "center-seeking!"

So now that you know the direction of an object's acceleration (toward the center of the circle), what about its magnitude? The formula for the magnitude of an object's centripetal acceleration is given by:

$$a_c = \frac{v^2}{r}$$

3.39 Q: In the diagram at right, a cart travels clockwise at constant speed in a horizontal circle. At the position shown in the diagram, which arrow indicates the direction of the centripetal acceleration of the cart?

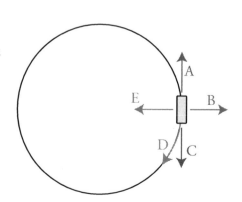

3.39 A: (E) The acceleration of any object moving in a circular path is toward the center of the circle.

3.40 Q: A 0.50-kilogram object moves in a horizontal circular path with a radius of 0.25 meter at a constant speed of 4.0 meters per second. What is the magnitude of the object's acceleration?

3.40 A: $a_c = \dfrac{v^2}{r} = \dfrac{\left(4\,^m\!/_s\right)^2}{0.25m} = 64\,^m\!/_{s^2}$

Circular Speed

So how do you find the speed of an object as it travels in a circular path? The formula for average speed that you learned previously applies.

$$\bar{v} = \frac{\Delta s}{t}$$

You have to be careful in using this equation, however, to understand that an object traveling in a circular path is traveling along the circumference of a circle. Therefore, if an object were to make one complete revolution around the circle, the distance it travels is equal to the circle's circumference.

$$C = 2\pi r$$

3.41 Q: Miranda drives her car clockwise around a circular track of diameter 60m. She completes 10 laps around the track in 2 minutes. Find Miranda's average speed and centripetal acceleration.

3.41 A: $\bar{v} = \dfrac{\Delta s}{t} = \dfrac{(10)(2\pi r)}{t} = \dfrac{(10)(2\pi)(30m)}{120s} = 15.7\,^m\!/_s$

$a_c = \dfrac{v^2}{r} = \dfrac{\left(15.7\,^m\!/_s\right)^2}{30m} = 8.2\,^m\!/_{s^2}$

Note that her displacement and average velocity are zero.

Relative Motion

You've probably heard the saying "motion is relative." Or perhaps you've heard people speak about Einstein's Theory of General Relativity and Einstein's Theory of Special Relativity. But what is this relativity concept?

In short, the concept of relative motion or relative velocity is all about understanding frame of reference. A frame of reference can be thought of as the state of motion of the observer of some event. For example, if you're sitting on a lawn chair watching a train travel past you from left to right at 50 m/s, you would consider yourself in a stationary frame of reference. From your perspective, you are at rest, and the train is moving. Further, assuming you have tremendous eyesight, you could even watch a glass of water sitting on a table inside the train move from left to right at 50 m/s.

An observer on the train itself, however, sitting beside the table with the glass of water, would view the glass of water as remaining stationary from their frame of reference. Because that observer is moving at 50 m/s, and the glass of water is moving at 50 m/s, the observer on the train sees no motion for the cup of water.

This seems like a simple and obvious example, yet when you take a step back and examine the bigger picture, you quickly find that all motion is relative. Going back to our original scenario, if you're sitting on your lawnchair watching a train go by, you believe you're in a stationary reference frame. The observer on the train looking out the window at you, however, sees you moving from right to left at 50 m/s.

Even more intriguing, an observer outside the Earth's atmosphere traveling with the Earth could use a "magic telescope" to observe you sitting in your lawnchair moving hundreds of meters per second as the Earth rotates about its axis. If this observer were further away from the Earth, he or she would also observe the Earth moving around the sun at speeds approaching 30,000 m/s. If the observer were even further away, they would observe the solar system (with the Earth, and you, on your lawnchair) orbiting the center of the Milky Way Galaxy at speeds approaching 220,000 m/s. And it goes on and on.

According to the laws of physics, there is no way to distinguish between an object at rest and an object moving at a constant velocity in an inertial (non-accelerating) reference frame. This means that there really is no "correct answer" to the question "how fast is the glass of water on the train moving?" You would be correct stating the glass is moving 50 m/s to the right and also correct in stating the glass is stationary. Imagine you're on a very smooth airplane, with all the window shades pulled down. It is physically impossible to determine whether you're flying through the air at a constant 300 m/s or whether you're sitting still on the runway. Even if you peeked out the window, you still couldn't say whether the plane was moving forward at 300 m/s, or the Earth was moving underneath the plane at 300 m/s.

As you observe, how fast you are moving depends upon the observer's frame of reference. This is what is meant by the statement "motion is relative." In order to determine an object's velocity, you really need to also state the reference frame (i.e. the train moves 50 m/s with respect to the ground; the glass of water moves 50 m/s with respect to the ground; the glass of water is stationary with respect to the train).

In most instances, the Earth makes a terrific frame of reference for physics problems. However, there are times when calculating the velocity of an object relative to different reference frames can be useful. Imagine you're in a canoe race, traveling down a river. It could be important to know not only your speed with respect to the flow of the river, but also your speed with respect to the riverbank, and even your speed with respect to your opponent's canoe in the race.

In dealing with these situations, you can state the velocity of an object with respect to its reference frame. For example, the velocity of object A with respect to reference frame C would be written as $\vec{v}_{AC}$. Even if you don't know the velocity of object A with respect to C directly, by finding the velocity of object A with respect to some intermediate object B, and the velocity of object B with respect to C, you can combine your velocities using vector addition to obtain:

$$\vec{v}_{AC} = \vec{v}_{AB} + \vec{v}_{BC}$$

This sounds more complicated than it actually is. Let's look at how this is applied in a few examples.

3.42 Q: A train travels at 60 m/s to the east with respect to the ground. A businessman on the train runs at 5 m/s to the west with respect to the train. Find the velocity of the man with respect to the ground.

3.42 A: First determine what information you are given. Calling east the positive direction, you know the velocity of the train with respect to the ground ($\vec{v}_{TG}$ =60 m/s). You also know the velocity of the man with respect to the train ($\vec{v}_{MT}$ =-5 m/s). Putting these together, you can find the velocity of the man with respect to the ground.
$$\vec{v}_{MG} = \vec{v}_{MT} + \vec{v}_{TG} = -5\,{}^m\!/_s + 60\,{}^m\!/_s = 55\,{}^m\!/_s$$

3.43 Q: An airplane flies at 250 m/s to the east with respect to the air. The air is moving at 15 m/s to the east with respect to the ground. Find the velocity of the plane with respect to the ground.

3.43 A: Again, start with the information you are given. If you call east positive, the velocity of the plane with respect to the air ($\vec{v}_{PA}$) is 250 m/s. The velocity of the air with respect to the ground ($\vec{v}_{AG}$) is 15 m/s. Solve for $\vec{v}_{PG}$.
$$\vec{v}_{PG} = \vec{v}_{PA} + \vec{v}_{AG} = 250\,{}^m\!/_s + 15\,{}^m\!/_s = 265\,{}^m\!/_s$$

This strategy isn't limited to one-dimensional problems. Treating velocities as vectors, you can use vector addition to solve problems in multiple dimensions.

3.44 Q: The president's airplane, Air Force One, flies at 250 m/s to the east with respect to the air. The air is moving at 35 m/s to the north with respect to the ground. Find the velocity of Air Force One with respect to the ground.

3.44 A: In this case, it's important to realize that both $\vec{v}_{PA}$ and $\vec{v}_{AG}$ are two-dimensional vectors. Once again, you can find $\vec{v}_{PG}$ by vector addition.

$$\vec{v}_{PG} = \vec{v}_{PA} + \vec{v}_{AG}$$

Drawing a diagram can be of tremendous assistance in solving this problem.

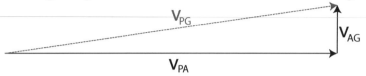

Looking at the diagram, you can easily solve for the magnitude of the velocity of the plane with respect to the ground using the Pythagorean Theorem.

$$v_{PG}^2 = v_{PA}^2 + v_{AG}^2 \rightarrow v_{PG} = \sqrt{v_{PA}^2 + v_{AG}^2} = \sqrt{(250\,^m/_s)^2 + (35\,^m/_s)^2} = 252\,^m/_s$$

You can find the angle of Air Force One using basic trig functions.

$$\tan\theta = \frac{opp}{adj} = \frac{v_{AG}}{v_{PA}} \rightarrow \theta = \tan^{-1}\left(\frac{v_{AG}}{v_{PA}}\right) = \tan^{-1}\left(\frac{35\,^m/_s}{250\,^m/_s}\right) = 8°$$

Therefore, the velocity of Air Force One with respect to the ground is 252 m/s at an angle of 8° north of east.

3.45 Q: A pilot wants to fly his Cessna 172 airplane from Seattle-Tacoma International Airport to Bremerton National Airport, located 18.4 miles away at a magnetic heading of nine degrees south of west. A 20 mile per hour wind is coming from the south.
A) If the Cessna's typical airspeed is 100 mph, about how long should it take the pilot to reach Bremerton Airport?
B) Determine the magnetic heading the pilot should follow in the air, taking into account the wind.

3.45 A: The pilot should follow the magnetic heading corresponding to the direction of the vector describing the velocity of the plane with respect to the air, $\vec{v}_{PA}$. The magnitude of this vector is given as 100 mph. You are also given the desired ground heading, which is the direction of the plane's desired velocity with respect to the ground ($\vec{v}_{PG}$). Finally, you are given the vector for the wind speed with respect to the ground. Keeping all units as miles per hour, you can start by writing the wind speed vector as $\vec{v}_{AG} = <0, 20>$.

You don't know the plane's speed with respect to the ground, but you can assign it the variable v. Therefore, the velocity of the plane with respect to the ground can be written in vector bracket notation as $\vec{v}_{PG} = <-v\cos 9°, -v\sin 9°>$.

You can then determine the velocity of the plane with respect to the air:

$$\vec{v}_{PA} = \vec{v}_{PG} + \vec{v}_{GA} = \vec{v}_{PG} - \vec{v}_{AG} \xrightarrow[\vec{v}_{AG}=<0,20>]{\vec{v}_{PG}=<-v\cos 9°, -v\sin 9°>} \vec{v}_{PA} = <-v\cos 9°, -v\sin 9° - 20>$$

Given the magnitude of $\vec{v}_{PA}$ is 100 mph, you can solve for v, the plane's speed with respect to the ground, using the Pythagorean Theorem.

$$|\vec{v}_{PA}| = 100 = \sqrt{(-v\cos 9°)^2 + (-v\sin 9° - 20)^2} \rightarrow v = 94.9\ mph$$

A) If the plane travels at 94.9 mph over a distance of 18.4 miles, you can find the approximate time for the flight.

$$\bar{v} = \frac{\Delta x}{t} \rightarrow t = \frac{\Delta x}{\bar{v}} = \frac{18.4\ miles}{94.9\ ^{miles}/_{hr}} = 0.194\ hrs = 11.6\ minutes$$

B) Determine the pilot's magnetic heading in the air by finding the direction of the $\vec{v}_{PA}$ vector.

$$\vec{v}_{PA} = <-v\cos 9°, -v\sin 9° - 20> \xrightarrow{v=94.9} \vec{v}_{PA} = <-93.7, -34.8>$$

The angle of this vector can be found from the tangent function:

$$\tan\theta = \frac{opp}{adj} \rightarrow \theta = \tan^{-1}\left(\frac{opp}{adj}\right) = \tan^{-1}\left(\frac{34.8}{93.7}\right) = 20.4°$$

Starting from Seattle, the pilot should fly a magnetic heading of 20.4 degrees south of west at an airspeed of 100 mph for 11.6 minutes to reach Bremerton National Airport.

3.46 Q: An eagle flies at constant velocity horizontally across the sky, carrying a turtle in its talons. The eagle releases the turtle while in flight. From the eagle's perspective, the turtle falls vertically with speed v_1. From an observer on the ground's perspective, at a particular instant the turtle falls at an angle with speed v_2. What is the speed of the eagle with respect to an observer on the ground?

A) $v_1 + v_2$

D) $\sqrt{v_2^2 - v_1^2}$

B) $v_1 - v_2$

E) $\sqrt{v_1^2 - 2v_1v_2 + v_2^2}$

C) $\sqrt{v_1^2 - v_2^2}$

3.46 A: (D) Define the velocity of the turtle with respect to the eagle $\vec{v}_{TE}$, which has magnitude v_1.
Define the velocity of the turtle with respect to the ground as $\vec{v}_{TG}$, with magnitude v_2.
You are asked to find the magnitude of the velocity of the eagle with respect to the ground ($|\vec{v}_{EG}|$). Making a diagram, you can use the Pythagorean Theorem to solve for the $|\vec{v}_{EG}|$.

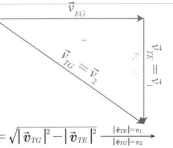

$$\vec{v}_{TG} = \vec{v}_{TE} + \vec{v}_{EG} \xrightarrow{Pythagorean\ Thrm} |\vec{v}_{TG}|^2 = |\vec{v}_{TE}|^2 + |\vec{v}_{EG}|^2 \rightarrow |\vec{v}_{EG}| = \sqrt{|\vec{v}_{TG}|^2 - |\vec{v}_{TE}|^2} \xrightarrow[|\dot{v}_{TG}|=v_2]{|\dot{v}_{TE}|=v_1}$$

$$|\vec{v}_{EG}| = \sqrt{v_2^2 - v_1^2}$$

3.47 Q: A car travels through a rainstorm at constant speed v_C as shown in the diagram at right. Rain is falling vertically at a constant speed v_R with respect to the ground. If the back windshield of the car, highlighted in the diagram, is set at an angle of θ with the vertical, what is the maximum speed the car can travel and still have rain hit the back windshield?

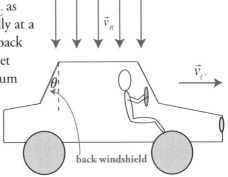

A) $v_R\cos\theta$

D) $v_R^2\sin\theta$

B) $v_R\tan\theta$

E) $\sqrt{v_R}\cos\theta$

C) $v_R\sin\theta$

back windshield

3.47 A: (B) In order for the rain to just hit the windshield, the angle of the velocity vector for the rain with respect to the car must match the angle of the back windshield. The velocity of the rain with respect to the car ($\vec{v}_{RC}$) can be found as the vector sum of the velocity of the rain with respect to the ground ($\vec{v}_R$) and the velocity of the ground with respect to the car ($-\vec{v}_C$) as shown in the diagram at right. From this diagram, it is a straightforward application of trigonometry to find the speed of the car at which this condition occurs.

$$\tan\theta = \frac{|\vec{v}_{GC}|}{|\vec{v}_{RG}|} = \frac{v_C}{v_R} \rightarrow v_C = v_R \tan\theta$$

Chapter 3: Kinematics

Chapter 4: Dynamics

"Use the force, Harry."

—Gandalf

Objectives

1. Define mass and inertia and explain the meaning of Newton's 1st Law.

2. Define a force as an interaction between two objects and identify corresponding force pairs.

3. Utilize free body diagrams (FBDs) to identify the forces exerted on an object.

4. Resolve vectors into perpendicular components to create pseudo-FBDs.

5. Write and solve Newton's 2nd Law equations corresponding to given FBDs or pseudo-FBDs.

6. Predict the motion of an object experiencing multiple forces by applying Newton's 2nd Law of Motion.

7. Define friction and distinguish between static and kinetic friction.

8. Determine the coefficient of friction for two surfaces.

9. Analyze bodies experiencing retarding forces which are a function of the body's velocity.

10. Calculate parallel and perpendicular components of an object's weight to solve ramp problems.

11. Identify force pairs in terms of Newton's 3rd Law of Motion.

12. Analyze and solve basic Atwood Machine problems using Newton's 2nd Law of Motion.

13. Identify centripetal forces and use them to solve problems involving objects undergoing circular motion.

Now that you've studied kinematics, you should have a pretty good understanding that objects in motion have **kinetic energy**, which is the ability of a moving object to move another object. To change an object's motion, and therefore its kinetic energy, the object must undergo a change in velocity, which is called an **acceleration**. So then, what causes an acceleration? To answer that question, you must study forces and their application.

Dynamics, or the study of forces, was very simply and effectively described by Sir Isaac Newton in 1686 in his masterpiece <u>Principia Mathematica Philosophiae Naturalis</u>. Newton described the relationship between forces and motion using three basic principles. Known as Newton's Laws of Motion, these concepts are still used today in applications ranging from sports science to aeronautical engineering.

Newton's 1st Law of Motion

Newton's 1st Law of Motion, also known as the **law of inertia**, can be summarized as follows:

> "An object at rest will remain at rest, and an object in motion will remain in motion, at constant velocity and in a straight line, unless acted upon by a net force."

This means that unless there is a net (unbalanced) force on an object, an object will continue in its current state of motion with a constant velocity. If this velocity is zero (the object is at rest), the object will continue to remain at rest. If this velocity is not zero, the object will continue to move in a straight line at the same speed. However, if a net (unbalanced) force does act on an object, that object's velocity will be changed (it will accelerate).

This sounds like a simple concept, but it can be quite confusing because it is difficult to observe in everyday life. People are usually fine with understanding the first part of the law: "an object at rest will remain at rest unless acted upon by a net force." This is easily observable. The donut sitting on your breakfast table this morning didn't spontaneously accelerate up into the sky. Nor did the family cat, Whiskers, lounging sleepily on the couch cushion the previous evening, all of a sudden accelerate sideways off the couch for no apparent reason.

The second part of the law contributes a considerably bigger challenge to the conceptual understanding of this principle. Realizing that "an object in motion will continue in its current state of motion with constant velocity unless acted upon by a net force" isn't easy to observe here on Earth, making this law rather tricky. Almost all objects observed in everyday life that are in motion are being acted upon by a net force - friction. Try this example: take your physics book and give it a good push along the floor. As expected, the book moves for some distance, but rather rapidly slides to a halt. An outside force, friction, has acted upon it. Therefore, from typical observations, it would be easy to think that an object must have a force continually applied upon it to remain in motion. However, this isn't so. If you took the same book out into the far reaches of space, away from any gravitational or frictional forces, and pushed it away, it would continue moving in a straight line at a constant velocity forever, as there are no external forces to change its motion. When the net force on an object is 0, the object is in **static equilibrium**. You'll revisit static equilibrium when discussing Newton's 2nd Law.

The tendency of an object to resist a change in velocity is known as the object's **inertia**. For example, a train has significantly more inertia than a skateboard. It is much harder to change the train's velocity than it is the skateboard's. The measure of an object's inertia is its **inertial mass**, typically referred to as mass. In other words, the more mass an object has, the greater its inertia.

4.1 Q: A 0.50-kilogram cart is rolling at a speed of 0.40 meter per second. If the speed of the cart is doubled, the inertia of the cart is
(A) quartered
(B) halved
(C) doubled
(D) quadrupled
(E) unchanged

4.1 A: (E) unchanged. Mass is a measure of an object's inertia, and the mass of the cart is constant.

4.2 Q: Which object has the greatest inertia?
(A) a 5.00-kg mass moving at 10.0 m/s
(B) a 10.0-kg mass moving at 1.00 m/s
(C) a 10.0-kg mass moving at 20 m/s
(D) a 15.0-kg mass moving at 10.0 m/s
(E) a 20.0-kg mass moving at 1.00 m/s

4.2 A: (E) a 20.0-kg mass has the greatest inertia.

If you recall from the kinematics unit, a change in velocity is known as an acceleration. Therefore, the second part of this law could be re-written to state that an object acted upon by a net force will be accelerated.

But what exactly is a force? **A force** is a vector quantity describing the push or pull on an object. Forces are measured in newtons (N), named after Sir Isaac Newton, of course. A newton is not a base unit, but is instead a derived unit, equivalent to 1 kg×m/s². Interestingly, the gravitational force on a medium-sized apple is approximately 1 newton.

You can break forces down into two basic types: contact forces and field forces. Contact forces occur when objects touch each other. Examples of contact forces include tension, friction, normal forces, elastic forces such as springs, and even the buoyant force. Field forces, also known as non-contact forces, occur at a distance. Examples of field forces include the gravitational force, the magnetic force, and the electrical force between two charged objects.

Interestingly, as you examine the universe on extremely small scales and on the most basic level, contact forces actually arise from interatomic electric field forces between objects. Therefore, viewed from a very basic perspective, all forces are ultimately field forces.

So, what then is a net force? A **net force** is just the vector sum of all the forces acting on an object. Imagine you and your sister are fighting over the last Christmas gift. You are pulling one end of the gift toward you with a force of 5N. Your sister is pulling the other end toward her (in the opposite direction) with a force of 5N. The net force on the gift, then, would be 0N; therefore, there would be no net force. As it turns out, though, you have a passion for Christmas gifts, and now increase your pulling force to 6N. The net force on the gift now is 1N in your direction; therefore, the gift would begin to accelerate toward you (yippee!). It can be difficult to keep track of all the forces acting on an object.

Free Body Diagrams

Fortunately, we have a terrific tool for analyzing the forces acting upon objects. This tool is known as a free body diagram (FBD). Quite simply, a **free body diagram** is a representation of a single object, or system, with vector arrows showing all the external forces acting on the object. These diagrams make it very easy to identify exactly what the net force is on an object, and they're also quite simple to create:

1. Isolate the object of interest. You can draw the object as a point or box representing the same mass. (When dealing with rotation later in this book, you'll have the added step of making sure the diagram represents the points on the object where the forces are applied.)
2. Sketch and label each of the external forces acting on the object, making sure to begin the force vectors at the point upon which the forces act.
3. Choose a coordinate system, with the direction of motion as one of the positive coordinate axes.
4. If all forces do not line up with your axes, resolve those forces into components using trigonometry.
5. Redraw your FBD, replacing forces that don't overlap the coordinates axes with their components to create a pseudo-FBD.

As an example, picture a glass of water sitting on the dining room table. You can represent the glass of water in the diagram as a rectangle. Then, represent each of the vector forces acting on the glass by drawing arrows and labeling them. In this case, you can start by recognizing the force of gravity on the glass, known more commonly as the glass's **weight**. Although you could label this force as F_{grav}, or W, get in the habit right now of writing the force of gravity on an object as mg. You can do this because the force of gravity on an object is equal to the object's mass times the acceleration due to gravity, g.

Of course, since the glass isn't accelerating, there must be another force acting on the glass to balance out the weight. This force, the force of the table pushing up on the glass, is known as the **normal force** (F_N). In physics, the normal force refers to a force perpendicular to a surface (normal in this case meaning perpendicular). The force of gravity on the glass must exactly match the normal force on the glass, although they are in opposite directions; therefore, there is no net force on the glass.

Chapter 4: Dynamics

4.3 Q: If the sum of all the forces acting on a moving object is zero, the object will
(A) slow down and stop
(B) change the direction of its motion
(C) accelerate uniformly
(D) continue moving with constant velocity
(E) turn in a circle at constant speed

4.3 A: (D) continue moving with constant velocity per Newton's 1st Law.

Newton's 2nd Law of Motion

Newton's 2nd Law of Motion may be the most important principle in all of modern-day physics because it explains exactly how an object's velocity is changed by a net force. In words, Newton's 2nd Law states that a net force applied to an object changes an object's velocity (produces an acceleration), and is frequently written as:

$$\sum \vec{F} = \vec{F}_{net} = m\vec{a} = m\frac{d\vec{v}}{dt}$$

It's important to remember that both force and acceleration are vectors. Therefore, the direction of the acceleration, or the change in center-of-mass velocity of an object, will be in the same direction as the net force. You can also look at this equation from the opposite perspective, stating that the acceleration of an object is directly proportional to the net force applied, and inversely proportional to the object's mass. In equation form:

$$\vec{a} = \frac{\vec{F}_{net}}{m}$$

You can analyze many situations involving both balanced and unbalanced forces on an object using the same basic steps.
1. Draw a free body diagram.
2. For any forces that don't line up with the x- or y-axes, break those forces up into components that do lie on the x- or y-axis.
3. Write expressions for the net force in x- and y- directions. Set the net force equal to ma, since Newton's 2nd Law tells us that $\vec{F}_{net} = m\vec{a}$.
4. Solve the resulting equations.

Let's take a look and see how these steps can be applied to a sample problem.

4.4 Q: A force of 25 newtons east and a force of 25 newtons west act concurrently on a 5-kilogram cart. Find the acceleration of the cart.
(A) 1.0 m/s² west
(B) 0.20 m/s² east
(C) 5.0 m/s² east
(D) 0 m/s²

4.4 A: Step 1: Draw a free-body diagram (FBD).
Step 2: All forces line up with x-axis. Define east as positive.
Steps 3&4: $\vec{F}_{net} = m\vec{a} \rightarrow 25N - 25N = m\vec{a} \rightarrow \vec{a} = 0$
Correct answer must be (D) 0 m/s².

Of course, everything you've already learned about kinematics still applies, and can be applied to dynamics problems as well. You'll find lots of application problems below, as this is a very important topic and common question type on the AP Physics C exam.

4.5 Q: A 0.15-kilogram baseball moving at 20 m/s is stopped by a catcher in 0.010 seconds. Find the average force stopping the ball.

4.5 A: Start by finding the acceleration of the ball as it is stopped by the catcher. Define the initial direction of the baseball as positive.

$$a = \frac{\Delta v}{t} = \frac{v - v_0}{t} = \frac{0 - 20\,m/s}{0.010s} = -2000\,m/s^2$$

The negative acceleration indicates the acceleration is in the direction opposite that of the initial velocity of the baseball. Now that you know acceleration, you can solve for force using Newton's 2nd Law.

$$F_{net} = ma = (0.15kg)(-2000\,m/s^2) = -300N$$

The negative sign in our answer indicates that the force applied is opposite the direction of the baseball's initial velocity.

4.6 Q: Three boxes travel without slipping along an assembly line moving at a constant speed of two meters per second to the right. If the boxes have masses of M, 2M, and 3M, respectively, which box will experience the greatest net force? Explain.

4.6 A: They all experience the same net force of zero because they are all moving at a constant velocity.

4.7 Q: Two forces, F_1 and F_2, are applied to an 8-kg block on a frictionless, horizontal surface as shown below.

$\vec{F}_1 = 12N$ Block $\vec{F}_2 = 2N$

Frictionless Surface

If the magnitude of the block's acceleration is 2.0 meters per second^2, what is the mass of the block?

4.7 A: Define left as the positive direction.

$$\vec{F}_{net} = m\vec{a} \rightarrow m = \frac{\vec{F}_{net}}{\vec{a}} = \frac{10N}{2\,m/s^2} = 5kg$$

4.8 Q: A 25-newton horizontal force northward and a 35-newton horizontal force southward act concurrently on a 15-kilogram object on a frictionless surface. What is the magnitude of the object's acceleration?
(A) 0.67 m/s²
(B) 1.7 m/s²
(C) 2.3 m/s²
(D) 4.0 m/s²

4.8 A: (A) $\vec{a} = \dfrac{\vec{F}_{net}}{m} = \dfrac{35N - 25N}{15kg} = 0.67\,{}^{m}\!/\!{}_{s^2}$

4.9 Q: A cardboard box of mass m on a wooden floor is represented by the free body diagram below.

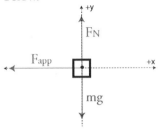

Given the FBD, which of the following expressions could be accurate mathematical representations for the box? Choose all that apply.
(A) $F_N - mg = ma_x$
(B) $-F_{app} = ma_x$
(C) $F_N - mg = ma_y$
(D) $F_{app} = ma_y$

4.9 A: (B) and (C): $-F_{app} = ma_x$ and $F_N - mg = ma_y$

4.10 Q: Three objects with differing masses are connected by strings and pulled by the right-most string with tension T_1 across a frictionless surface as shown in the diagram below.

Which of the following expressions accurately depicts the acceleration of the system? Choose all that apply.
(A) T3/2m
(B) T2/3m
(C) T1/6m
(D) None of the above.

4.10 A: Both (B) and (C) are correct.
Writing Newton's 2nd Law equations in the x-direction for each of the three objects:
$$\vec{F}_{net_x} = m\vec{a}_x$$
$$\vec{T}_3 = m\vec{a} \rightarrow \vec{a} = \frac{\vec{T}_3}{m}$$
$$\vec{T}_2 - \vec{T}_3 = 2m\vec{a} \xrightarrow[\vec{T}_2 = 3m\vec{a}]{\vec{T}_3 = m\vec{a}} \vec{a} = \frac{\vec{T}_2}{3m}$$
$$\vec{T}_1 - \vec{T}_2 = 3m\vec{a} \xrightarrow[\vec{T}_1 = 6m\vec{a}]{\vec{T}_2 = 3m\vec{a}} \vec{a} = \frac{\vec{T}_1}{6m}$$

4.11 Q: What is the net force on an object experiencing a pull of 5N to the north, a push of 3N to the south, and a pull of 2N to the east?

4.11 A: Start with a free body diagram, as shown below left. Then, line up the force vectors tip-to-tail and draw a line from the starting point of the first vector to the ending point of the last vector in order to determine the resultant, shown below right.

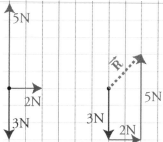

You can then determine the magnitude and direction of the resultant using the Pythagorean Theorem: $|\vec{F}_{net}| = |\vec{R}| = \sqrt{(2N)^2 + (2N)^2} = 2.83N$ at $45°$ *North of East*.

4.12 Q: A hummingbird spontaneously begins humming the "Nae Nae" song, causing its companion hummingbird of mass 4 grams to immediately fly away in a straight line with velocity vector $\vec{v}(t) = (3t^3 + 2)\hat{i}$. Determine the net force on the escaping hummingbird at time t=3 seconds.

4.12 A: $\vec{F}_{net} = m\dfrac{d\vec{v}}{dt} = 0.004\dfrac{d}{dt}(3t^3 + 2)\hat{i} = 0.004(9t^2)\hat{i} = 0.036t^2\hat{i} \xrightarrow{t=3s} \vec{F}_{net} = (0.324N)\hat{i}$

4.13 Q: A particle follows the path from point 1 to point 2 as shown in the diagrams below.

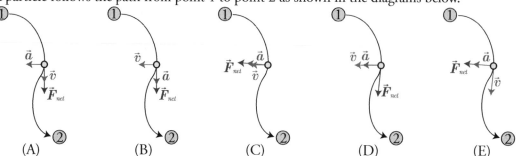

Which diagram best depicts possible velocity, acceleration, and net force vectors for the particle at the given position?

4.13 A: (E) is the best choice, as the acceleration and net force vectors are in the same direction, and the velocity is tangent to the path traveled.

Chapter 4: Dynamics

4.14 Q: An astronaut weighs 1000N on Earth.
A) What is the weight of the astronaut on Planet X, where the gravitational field strength (g) is 6 m/s²?
B) An alien on Planet X weighs 400N. What is the mass of the alien?

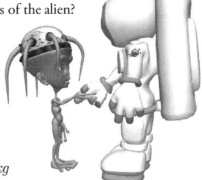

4.14 A: A) First find the astronaut's mass on Earth.

$$mg_{Earth} = 1000N \rightarrow m = \frac{1000N}{g_{Earth}} = \frac{1000N}{10\,{}^m\!/_{s^2}} = 100kg$$

Next, determine the astronaut's weight on Planet X.

$$mg_X = (100kg)\,(6\,{}^m\!/_{s^2}) = 600N$$

B) $m_{Alien}\,g_X = 400N \rightarrow m_{Alien} = \dfrac{400N}{g_X} = \dfrac{400N}{6\,{}^m\!/_{s^2}} = 66.7kg$

4.15 Q: A 15-kg wagon is pulled to the right across a surface by a tension of 100 newtons at an angle of 30 degrees above the horizontal. A frictional force of 20 newtons to the left acts simultaneously. What is the acceleration of the wagon?

4.15 A: First, draw a diagram of the situation.

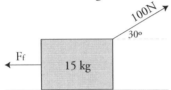

Next, create a FBD and pseudo-FBD.

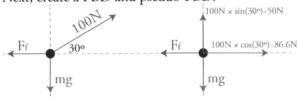

Finally, use Newton's 2nd Law in the x-direction to determine the horizontal acceleration of the wagon.

$$\vec{F}_{net_x} = m\vec{a}_x \rightarrow \vec{a}_x = \frac{\vec{F}_{net_x}}{m} = \frac{(86.6N - 20N)\,\hat{i}}{15kg} = 4.44\,{}^m\!/_{s^2}\hat{i}$$

4.16 Q: A steel beam of mass 100 kg is attached by a cable to a 200 kg beam above it. The two are raised upward with an acceleration of 1 m/s² by another cable attached to the top beam. Assuming the cables are massless, find the tension in each cable.

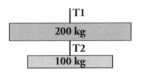

4.16 A: Begin by drawing free body diagrams for each of the beams.

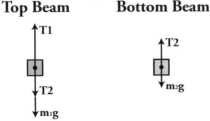

Next, you can write a Newton's 2nd Law equation for each free body diagram to solve for the unknown tensions. Beginning with Beam 2:

$$\vec{F}_{net_y} = \vec{T}_2 - m_2\vec{g} = m_2\vec{a}_y \rightarrow \vec{T}_2 = m_2\left(\vec{g} + \vec{a}_y\right) = (100kg)\left(10^{m}/_{s^2}\hat{j} + 1^{m}/_{s^2}\hat{j}\right) = 1100N\hat{j}$$

Now you can solve for T_1 by writing a Newton's 2nd Law equation for Beam 1.

$$\vec{F}_{net_y} = \vec{T}_1 - \vec{T}_2 - m_1\vec{g} = m_1\vec{a}_y \rightarrow \vec{T}_1 = \vec{T}_2 + m_1\left(\vec{g} + \vec{a}_y\right) \xrightarrow[g=10^{m}/_{s^2},\ a=1^{m}/_{s^2}]{T_2=1100N} \vec{T}_1 = 3300N\hat{j}$$

4.17 Q: The system below is accelerated by applying a tension T_1 to the right-most cable. Assuming the system is frictionless, determine the tension T_2 in terms of T_1.

| 2 kg | —T2— | 5 kg | T1→ |

4.17 A: As T_1 pulls to the right, it accelerates a total mass of 7 kg with an acceleration $\vec{a}$:

$$\vec{F}_{net_x} = m\vec{a}_x \rightarrow \vec{a}_x = \frac{\vec{F}_{net_x}}{m} = \frac{\vec{T}_1}{7kg}$$

T_2 only accelerates the 2-kg mass with acceleration $\vec{a}$, therefore you can utilize Newton's 2nd Law along with your previous expression for $\vec{a}_x$ to obtain T_2 in terms of T_1.

$$\vec{F}_{net_x} = m\vec{a}_x \rightarrow \vec{T}_2 = (2kg)\,\vec{a}_x \xrightarrow{\vec{a}_x=\frac{\vec{T}_1}{7kg}} \vec{T}_2 = (2kg)\frac{\vec{T}_1}{7kg} = \frac{2}{7}\vec{T}_1$$

4.18 Q: The velocity of an airplane in flight is shown as a function of time t in the graph. At which time is the force exerted by the plane's engines on the plane the greatest?

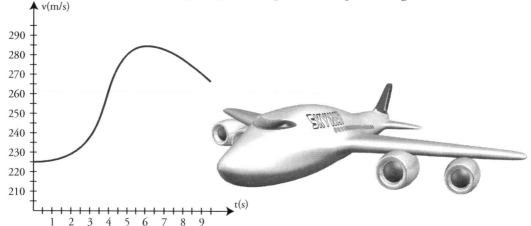

4.18 A: The force is greatest where the acceleration is greatest, which occurs when the v-t graph has the steepest slope at approximately t=4 seconds according to the graph.

Chapter 4: Dynamics

Elevators

Elevators make great demonstration vehicles for exploring Newton's Laws of Motion, and are a popular topic on standardized physics exams. Typically, the demonstration involves a person standing on a scale in an elevator (for reasons unknown to most physics instructors).

In analyzing elevator problems, it is important first to realize that a scale does not directly read the weight of the person upon it. Rather, the scale provides a measure of the normal force the scale exerts. Once you realize that the reading on a scale is the normal force, analyzing problems involving elevators becomes a straightforward exercise in applying Newton's Laws of Motion.

4.19 Q: A man with mass m stands on a scale in an elevator. If the scale reading is equal to mg when the elevator is at rest, what is the scale reading while the elevator is accelerating downwards with a magnitude of a?
(A) $m(g-a_y)$
(B) $m(g+a_y)$
(C) $m(a_y-g)$
(D) $m(a_y-g)$

4.19 A: Since the elevator is moving downward, set the positive y axis as pointing downward. You can then draw a free body diagram for the man in the elevator and apply Newton's 2nd Law in the y-direction.

$$\vec{F}_{net_y} = m\vec{a}_y \rightarrow m\vec{g} - \vec{F}_N = m\vec{a}_y \rightarrow \vec{F}_N = m\vec{g} - m\vec{a}_y \rightarrow \vec{F}_N = m(\vec{g} - \vec{a}_y)$$

Therefore the correct answer is (A) $m(g-a_y)$.

4.20 Q: Lizzie stands on a scale in an elevator. If the scale on the elevator reads 600N when Lizzie is riding upward at a constant 4 m/s, what is the reading on the scale when the elevator is at rest?
(A) 420 N
(B) 600 N
(C) 780 N
(D) 840 N

4.20 A: (B) The scale reads the exact same at rest as it does while moving at constant velocity.

Equilibrium

The special situation in which the net force on an object turns out to be zero, called **equilibrium**, tells you immediately that the object isn't accelerating. If the object is moving with some velocity, it will remain moving with that exact same velocity known as **dynamic equilibrium**. If the object is at rest, it will remain at rest, known as **static equilibrium**. Sounds familiar, doesn't it? This is a restatement of Newton's 1st Law of Motion, the Law of Inertia. Consider the situation of a tug-of-war. If both participants are pulling with tremendous force, but the force is balanced, there is no acceleration — a great example of equilibrium.

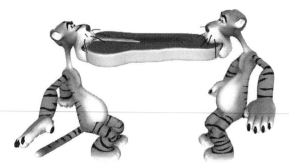

Equilibrium conditions are so widespread that knowing how to explore and analyze these conditions is a key stepping stone to understanding more complex situations.

One common analysis question involves finding the equilibrant force given a free body diagram of an object. The **equilibrant** is a single force vector that you add to the unbalanced forces on an object in order to bring it into static equilibrium. For example, if you are given a force vector of 10N north and 10N east, and are asked to find the equilibrant, you're really being asked to find a force that will offset the two given forces, bringing the object into static equilibrium.

To find the equilibrant, you must first find the net force being applied to the object. To do this, apply your vector math and add up the two vectors by first lining them up tip to tail, then drawing a straight line from the starting point of the first vector to the ending point of the last vector. The magnitude of this vector can be found using the Pythagorean Theorem.

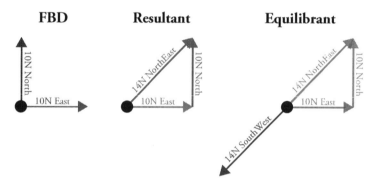

Finally, to find the equilibrant vector, add a single vector to the diagram that will give a net force of zero. If your total net force is currently 14N northeast, then the vector that should bring this back into equilibrium, the equilibrant, must be the opposite of 14N northeast, or a vector with magnitude 14N to the southwest.

4.21 Q: A 20-newton force due north and a 20-newton force due east act concurrently on an object, as shown in the diagram below.

The additional force necessary to bring the object into a state of equilibrium is
(A) 40 N northeast
(B) 20 N southwest
(C) 28 N northeast
(D) 28 N southwest
(E) 40 N southwest

4.21 A: (D) The resultant vector is 28 newtons northeast, so its equilibrant must be 28 newtons southwest.

Another common analysis question involves asking whether three vectors could be arranged to provide a static equilibrium situation.

4.22 Q: A 3-newton force and a 4-newton force are acting concurrently on a point. Which force could not produce equilibrium with these two forces?
(A) 1 N
(B) 2 N
(C) 4 N
(D) 7 N
(E) 9 N

4.22 A: (E) A 9-newton force could not produce equilibrium with a 3-newton and a 4-newton force. One way to test this is to draw vectors of the three forces. If you can arrange the vectors to create a closed triangle, they can produce equilibrium.

4.23 Q: Determine the tensions T_1 and T_2 in cables holding a 10-newton weight stationary.

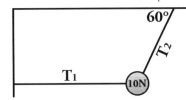

4.23 A: Start with a free body diagram, then break the force at an angle (T_2) into its components to build a pseudo-FBD.

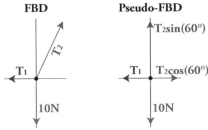

Next you can apply Newton's 2nd Law in order to determine the tension in each cable.

$$F_{net_y} = T_2 \sin 60° - 10N = ma_y \xrightarrow{a_y=0} T_2 = \frac{10N}{\sin 60°} = 11.5N$$

$$F_{net_x} = T_2 \cos 60° - T_1 = ma_x \xrightarrow[T_2=11.5N]{a_x=0} T_1 = (11.5N) \cos 60° = 5.77N$$

4.24 Q: A traffic light is suspended by two cables as shown in the diagram. If cable 1 has a tension T_1=50 newtons, and cable 2 has a tension T_2=86.6 newtons, find the mass of the traffic light.

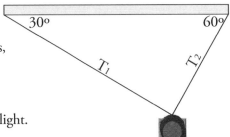

4.24 A: First draw the FBD and pseudo-FBD for the traffic light.

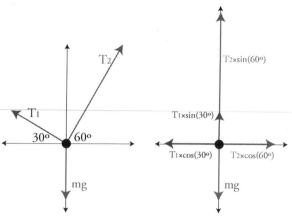

Now, you can write a Newton's 2nd Law Equation using the pseudo-FBD and solve for the mass of the stoplight.

$$\vec{F}_{nety} = T_1\sin 30° + T_2\sin 60° - mg = ma_y \xrightarrow{a_y=0} m = \frac{T_1\sin 30° + T_2\sin 60°}{g} \rightarrow$$

$$m = \frac{50N\sin 30° + 86.6N\sin 60°}{10^m/_{s^2}} = 10kg$$

Friction

Up until this point, it's been convenient to ignore one of the most useful and most troublesome forces in everyday life... a force that has tremendous application in transportation, machinery, and all parts of mechanics, yet people spend tremendous amounts of effort and money each day fighting it. This force, **friction**, is a resistive force to the relative motion between two surfaces in contact. Friction opposes motion.

4.25 Q: A projectile launched at an angle of 45° above the horizontal travels through the air. Compared to the projectile's theoretical path with no air friction, the actual trajectory of the projectile with air friction is
(A) lower and shorter
(B) lower and longer
(C) higher and shorter
(D) higher and longer

4.25 A: (A) lower and shorter. Friction opposes motion.

Author's Note: As you progress deeper into your study of physics, you may find substantiated arguments against referring to air resistance as a form of friction. For the purposes of simplicity in this course, it is reasonable to assume air friction and air resistance are the same thing. Just be aware that in future studies these definitions may require further refinement.

4.26 Q: A box is pushed toward the right across a classroom floor. The force of friction on the box is directed toward the
(A) left
(B) right
(C) ceiling
(D) floor

4.26 A: (A) left. Friction opposes motion.

There are two main types of friction. **Kinetic friction** is a frictional force that opposes motion for an object which is sliding along another surface. **Static friction**, on the other hand, acts on an object that isn't sliding. If you push on your textbook, but not so hard that it slides along your desk, static friction is opposing your applied force on the book, leaving the book in static equilibrium.

The magnitude of the frictional force depends upon two factors:
1. The nature of the surfaces in contact.
2. The normal force acting on the object (F_N).

The ratio of the frictional force and the normal force provides the **coefficient of friction** (μ), a proportionality constant that is specific to the two materials in contact.

You can look up the coefficient of friction for various surfaces. Make sure you choose the appropriate coefficient. Use the static coefficient (μ_s) for objects which are not sliding, and the kinetic coefficient (μ_k) for objects which are sliding.

Approximate Coefficients of Friction	Kinetic	Static
Rubber on concrete (dry)	0.68	0.90
Rubber on concrete (wet)	0.58	
Rubber on asphalt (dry)	0.67	0.85
Rubber on asphalt (wet)	0.53	
Rubber on ice	0.15	
Waxed ski on snow	0.05	0.14
Wood on wood	0.30	0.42
Steel on steel	0.57	0.74
Copper on steel	0.36	0.53
Teflon on Teflon	0.04	

Which coefficient would you use for a sled sliding down a snowy hill? The kinetic coefficient of friction, of course. How about a refrigerator on your linoleum floor that is at rest and you want to start in motion? That would be the static coefficient of friction. Let's try a harder one: a car drives with its tires rolling freely. Is the friction between the tires and the road static or kinetic? Static. The tires are in constant contact with the road, much like walking. If the car was skidding, however, and the tires were locked, you would look at kinetic friction. Let's take a look at some sample problems:

4.27 Q: A car's performance is tested on various horizontal road surfaces. The brakes are applied, causing the rubber tires of the car to slide along the road without rolling. The tires encounter the greatest force of friction to stop the car on
(A) dry concrete
(B) dry asphalt
(C) wet concrete
(D) wet asphalt

4.27 A: To obtain the greatest force of friction (F_f), you'll need the greatest coefficient of friction (μ). Use the kinetic coefficient of friction (μ_k) since the tires are sliding. From the Approximate Coefficients of Friction table, the highest kinetic coefficient of friction for rubber comes from rubber on dry concrete. Answer: (A).

4.28 Q: The diagram below shows a block sliding down a plane inclined at angle θ with the horizontal.

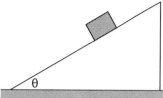

As angle θ is increased, the coefficient of kinetic friction between the bottom surface of the block and the surface of the incline will
(A) decrease
(B) increase
(C) remain the same

4.28 A: (C) remain the same. Coefficient of friction depends only upon the materials in contact.

The normal force always acts perpendicular to a surface, and comes from the interaction between atoms that act to maintain its shape. In many cases, it can be thought of as the elastic force trying to keep a flat surface flat (instead of bowed). You can use the normal force to calculate the magnitude of the frictional force.

The force of friction, depending only upon the nature of the surfaces in contact (μ) and the magnitude of the normal force (F_N), can be determined using the formula:
$$|\vec{F}_f| \le \mu |\vec{F}_N|$$

It is important to note that the frictional force will match the applied force until the applied force overcomes the maximum frictional force, which occurs when the magnitude of the frictional force is equal to the coefficient of friction multiplied by the magnitude of the normal force ($F_f = \mu F_N$).

Solving problems involving friction requires application of the same basic principles you've been learning about throughout the dynamics unit: drawing a free body diagram, applying Newton's 2nd Law along the x- and/or y-axes, and solving for any unknowns. The only new skill is drawing the frictional force on the free body diagram, and using the relationship between the force of friction and the normal force to solve for the unknowns. Let's take a look at some more sample problems:

4.29 Q: The diagram below shows a 10.0-kilogram object accelerating at 2 meters per second^2 on a rough horizontal surface. Determine the magnitude of F_f.

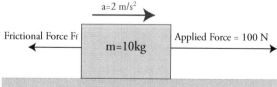

4.29 A: Define to the right as the positive direction.

$$\vec{F}_{net_x} = m\vec{a}_x \rightarrow \vec{F}_{app} - \vec{F}_f = m\vec{a}_x \rightarrow \vec{F}_f = \vec{F}_{app} - m\vec{a}_x = 100N - (10kg)(2^m/_{s^2}) = 80N$$

4.30 Q: An ice skater applies a horizontal force to a 20-kilogram block on frictionless, level ice, causing the block to accelerate uniformly at 1.4 m/s² to the right. After the skater stops pushing the block, it slides onto a region of ice that is covered by a thin layer of sand. The coefficient of kinetic friction between the block and the sand-covered ice is 0.28.
A) Calculate the magnitude of the force applied to the block by the skater.
B) Determine the magnitude of the normal force acting on the block.
C) Calculate the magnitude of the force of friction acting on the block as it slides over the sand-covered ice.

4.30 A: A) Define right as the positive direction.

$$\vec{F}_{net_x} = m\vec{a}_x = (20kg)(1.4^m/_{s^2}) = 28N$$

B) $\vec{F}_{net_y} = m\vec{a}_y \xrightarrow{\vec{a}_y = 0} F_N - mg = 0 \rightarrow$

$$F_N = mg = (20kg)(10^m/_{s^2}) = 200N$$

C) $|\vec{F}_f| = \mu|\vec{F}_N| = (0.28)(200N) = 56N$

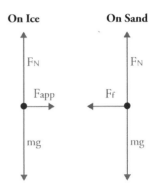

4.31 Q: Johnny pulls the handle of a 20-kilogram sled across the yard with a force of 100N at an angle of 30 degrees above the horizontal. The yard exerts a frictional force of 25N on the sled.
A) Find the coefficient of friction between the sled and the yard.
B) Determine the distance the sled travels if it starts from rest and Johnny maintains his 100N force for five seconds.

4.31 A: A) First draw a FBD and pseudo-FBD for the situation.

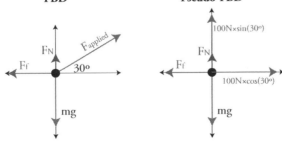

Next, write a Newton's 2nd Law Equation in the y-direction to solve for the normal force.

$$F_{net_y} = 100N\sin 30° + F_N - mg = ma_y \xrightarrow{a_y = 0} F_N = 200N - 50N = 150N$$

Then, find the coefficient of friction by dividing the frictional force by the normal force.

$$\mu = \frac{|\vec{F}_f|}{|\vec{F}_N|} = \frac{25N}{150N} = 0.167$$

B) In order to determine the distance the sled travels if it starts from rest, first find the acceleration of the sled by writing a Newton's 2nd Law Equation in the x-direction and solving for the horizontal acceleration.

$$F_{net_x} = 100N \cos 30° - F_f = 86.6N - 25N = ma_x \rightarrow ma_x = \frac{61.6N}{20kg} = 3.1\,{}^m\!/_{s^2}$$

Finally, use the kinematic equations to find the distance the sled travels.
$$\Delta x = v_{x_0}t + \tfrac{1}{2}a_x t^2 = \tfrac{1}{2}(3.1\,{}^m\!/_{s^2})(5s)^2 = 38.8m$$

4.32 Q: A 10-kg wooden box sits on a wooden surface. The coefficient of static friction between the two surfaces is 0.42. A horizontal force of 20N is applied to the box. What is the magnitude of the force of friction on the box?

4.32 A: Start by finding the maximum force of friction on the box.
$$|\vec{F}_f| \le \mu |\vec{F}_N| \xrightarrow[m=10kg]{F_N=mg} |\vec{F}_f| \le (0.42)(10kg)(10\,{}^m\!/_{s^2}) \rightarrow |\vec{F}_f| \le 42N$$

The force of friction will match the applied force on the box to keep it at rest and in static equilibrium as long as the applied force remains less than 42N. Therefore, the force of friction on the box must be 20N.

Retarding and Drag Forces

Throughout this course, a common source of friction observed in everyday life is often neglected. **Air resistance**, also known as drag or fluid resistance, is a form of fluid friction between a moving object and the air around it, and can play a significant role as a force affecting moving bodies. Modeling air resistance mathematically can be a detailed and involved process because the force of air resistance depends on the object's velocity.

When an object falls through the atmosphere, it accelerates until it reaches its maximum vertical velocity, known as its **terminal velocity** (v_t). Once an object reaches its terminal velocity, it continues to fall at constant velocity with no acceleration. In this situation, the force of gravity on the object balances the drag force on the object. With zero net force, the object is in dynamic equilibrium, and continues to fall at a constant velocity as indicated by Newton's 1st Law of Motion.

Assume David falls from an airplane. Typically, the drag forces on a free-falling object take a form similar to $F_{drag} = bv$ or $F_{drag} = cv^2$, where b and c are constants. For the sake of this problem, let's assume $F_{drag} = bv$. You could then write a Newton's 2nd Law Equation in the y-direction for the skydiver as $\vec{F}_{net_y} = m\vec{g} - \vec{F}_{drag} = m\vec{a}_y \xrightarrow{F_{drag}=bv} mg - bv = ma_y$.

Initially, at time t=0, velocity=0, therefore a=g. After a long time, however, David reaches terminal velocity (v_t) and a=0. At this point, $F_{drag} = mg$, allowing you to refine the previous equation in order to solve for the terminal velocity: $mg - bv = ma_y \xrightarrow[v=v_t]{a_y=0} mg - bv_t = 0 \rightarrow v_t = \frac{mg}{b}$.

You can go back to the Newton's 2nd Law equation to find velocity itself as a function of time. This is actually a differential equation, in which velocity and its derivative with respect to time, acceleration, are both in the same equation. You can solve this by using a technique known as separation of variables.

Begin by re-writing the Newton's 2nd Law equation in the vertical direction, directly expressing acceleration as the derivative of velocity with respect to time. This makes it easier to determine which expressions are related to which variable.

$$mg - bv = ma_y \xrightarrow{a_y = \frac{dv}{dt}} mg - bv = m\frac{dv}{dt}$$

Next, rearrange the equation algebraically so that all the expressions relying on the variable velocity are on one side of the equation, and all the expressions relying on the variable time are on the other side of the equation. While rearranging the expressions, you may simplify the expression by replacing mg/b with the terminal velocity v_t.

$$mg - bv = m\frac{dv}{dt} \rightarrow \frac{mg}{b} - v = \frac{m}{b}\frac{dv}{dt} \xrightarrow{v_t = \frac{mg}{b}} v_t - v = \frac{m}{b}\frac{dv}{dt} \rightarrow$$

$$\frac{dv}{v_t - v} = \frac{b}{m}dt \rightarrow \frac{dv}{v - v_t} = -\frac{b}{m}dt$$

Now, with the velocity variables separated from the time variables, you can integrate both sides.

$$\int_0^v \frac{dv}{v - v_t} = \int_0^t -\frac{b}{m}dt$$

The left-hand integral can be integrated easily by recognizing the expression fits the form of the natural logarithm integral. The right-hand side is a straightforward integral of a constant with respect to t. Integrate as follows:

$$\int_0^v \frac{dv}{v - v_t} = \int_0^t -\frac{b}{m}dt \xrightarrow[\substack{du = dv \\ \int \frac{du}{u} = \ln u + C}]{u = v - v_t} \ln(v - v_t) \Big|_0^v = -\frac{b}{m}t \Big|_0^t$$

Next you can solve for velocity as a function of time using a series of algebraic manipulations.

$$\ln(v - v_t) \Big|_0^v = -\frac{b}{m}t \Big|_0^t \rightarrow \ln(v - v_t) - \ln(-v_t) = -\frac{b}{m}t \xrightarrow{\ln a - \ln b = \ln \frac{a}{b}} \ln\left(\frac{v - v_t}{-v_t}\right) = -\frac{b}{m}t \rightarrow$$

$$\ln\left(\frac{v_t - v}{v_t}\right) = -\frac{b}{m}t \rightarrow \ln\left(1 - \frac{v}{v_t}\right) = -\frac{b}{m}t \rightarrow e^{\ln\left(1 - \frac{v}{v_t}\right)} = e^{-\frac{b}{m}t} \rightarrow 1 - \frac{v}{v_t} = e^{-\frac{b}{m}t} \rightarrow \frac{v}{v_t} = 1 - e^{-\frac{b}{m}t} \rightarrow$$

$$v = v_t\left(1 - e^{-\frac{b}{m}t}\right)$$

Quite a series of steps, but all algebra, working to isolate velocity on the left. Note that if you hadn't replaced mg/b with the terminal velocity v_t, you would have obtained the equivalent equation:

$$v = \frac{mg}{b}\left(1 - e^{-\frac{b}{m}t}\right)$$

Now that you've solved for the velocity as a function of time, you can solve for the acceleration as a function of time by taking the derivative of the velocity with respect to time.

$$v = v_t\left(1 - e^{-\frac{b}{m}t}\right) = v_t - v_t e^{-\frac{b}{m}t} \xrightarrow{a = \frac{dv}{dt}} a = \frac{d}{dt}\left(v_t - v_t e^{-\frac{b}{m}t}\right) = \frac{d}{dt}(v_t) - \frac{d}{dt}\left(v_t e^{-\frac{b}{m}t}\right)$$

Recognizing v_t is actually a constant, the calculation for acceleration becomes straightforward.

$$a = \frac{d}{dt}(v_t) - \frac{d}{dt}\left(v_t e^{-\frac{b}{m}t}\right) = -v_t\frac{d}{dt}\left(e^{-\frac{b}{m}t}\right) \xrightarrow{\frac{d}{dx}(e^{au}) = ae^{au}} a = -v_t\left(-\frac{b}{m}\right)e^{-\frac{b}{m}t}$$

Further simplifications are available by replacing the terminal velocity with mg/b:

$$a = -v_t\left(-\frac{b}{m}\right)e^{-\frac{b}{m}t} \xrightarrow{v_t = \frac{mg}{b}} a = -\left(\frac{mg}{b}\right)\left(-\frac{b}{m}\right)e^{-\frac{b}{m}t} \rightarrow a = ge^{-\frac{b}{m}t}$$

Hint: Note the form of the solution. For these types of problems, you will always see something of the form $1 - e^{-\frac{t}{\tau}}$ or $e^{-\frac{t}{\tau}}$, where τ is some time constant. You can use the shape of the graph at t=0 and as t approaches infinity to tell you which form to use. It is often advantageous to guess at the form of the solution using what you know about the problem, prior to diving into a full solution.

Let's examine the position vs. time, velocity vs. time, and acceleration vs. time graphs to provide a better picture of how drag forces affect a falling object. Note that all graphs are drawn assuming down as the positive y-axis direction.

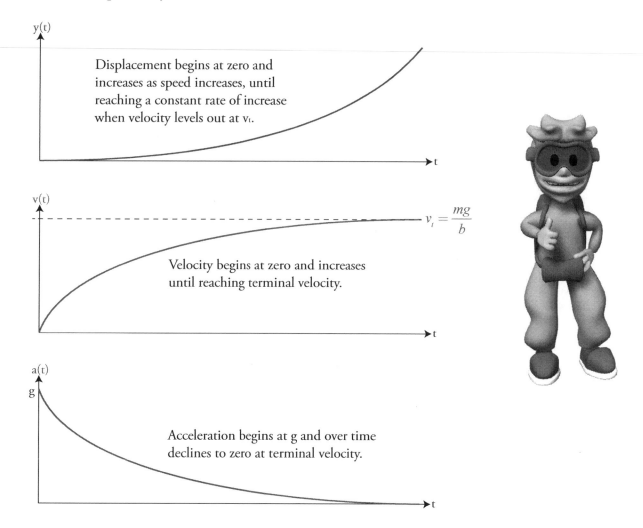

y(t)

Displacement begins at zero and increases as speed increases, until reaching a constant rate of increase when velocity levels out at v_t.

v(t)

$v_t = \dfrac{mg}{b}$

Velocity begins at zero and increases until reaching terminal velocity.

a(t)

g

Acceleration begins at g and over time declines to zero at terminal velocity.

4.33 Q: After talking all through the latest superhero movie, fellow movie goers push 100-kg Leroy off a vertical cliff where he falls with a drag force due to air resistance modeled as $F_{drag} = cv^2$. If Leroy achieves a terminal velocity of 50 m/s, what is the value of the drag coefficient c?

4.33 A: Upon reaching terminal velocity, the force of gravity on Leroy is equal in magnitude to the drag force on relay, therefore you can use Newton's 2nd Law to write:

$$F_{net_y} = mg - cv^2 = ma_y \xrightarrow[a_y=0]{v=v_t} mg - cv_t^2 = 0 \rightarrow c = \frac{mg}{v_t^2}$$

Then, solve for c by substituting in your given information.

$$c = \frac{mg}{v_t^2} = \frac{(100kg)\left(10^{m}/_{s^2}\right)}{\left(50^{m}/_{s}\right)^2} = 0.4^{kg}/_{m}$$

4.34 Q: A ball is dropped from rest at time t=0 and experiences air resistance such that its acceleration is given by a=g-kv. Determine the acceleration of the ball as a function of time.

4.34 A: Beginning with the given equation, you can rewrite acceleration as the derivative of velocity with respect to time.

$$a = g - kv \xrightarrow{a=\frac{dv}{dt}} \frac{dv}{dt} = g - kv$$

Next, apply the separation of variables method and integrate both sides of the equation.

$$\frac{dv}{dt} = g - kv \rightarrow \frac{dv}{g - kv} = dt \rightarrow \int_0^v \frac{dv}{g - kv} = \int_0^t dt$$

Now you can solve the resulting equation to obtain the velocity as a function of time.

$$\int_0^v \frac{dv}{g - kv} = \int_0^t dt \xrightarrow[\int \frac{du}{u} = \ln u + C]{\substack{u=g-kv \\ du=-kdv}} -\frac{1}{k}\ln(g-kv)\Big|_0^v = t\Big|_0^t \rightarrow \ln(g-kv) - \ln(g) = -kt \rightarrow$$

$$\ln\left(\frac{g-kv}{g}\right) = -kt \rightarrow e^{\ln\left(\frac{g-kv}{g}\right)} = e^{-kt} \rightarrow \frac{g-kv}{g} = e^{-kt} \rightarrow g - kv = ge^{-kt} \rightarrow kv = g - ge^{-kt} \rightarrow$$

$$kv = g(1 - e^{-kt}) \rightarrow v = \frac{g}{k}(1 - e^{-kt})$$

Now that you have velocity as a function of time, find acceleration by taking the derivative of velocity with respect to time.

$$a = \frac{dv}{dt} = \frac{d}{dt}\left(\frac{g}{k}(1 - e^{-kt})\right) = \frac{g}{k}\frac{d}{dt}(1 - e^{-kt}) \xrightarrow{\frac{d}{dx}(e^{au})=ae^{au}} a = -\frac{g}{k}(-k)e^{-kt} \rightarrow a(t) = ge^{-kt}$$

4.35 Q: A submarine of mass m travels horizontally through the ocean with speed v_0. At time t=0 it turns off its engines to coast silently, encountering a fluid resistive force with magnitude cv^2, where c is a constant.
(A) Determine the speed of the submarine as a function of time.
(B) Determine the speed of the submarine at time t=2 minutes if the submarine has a mass of 10^7 kg, an initial speed of 20 m/s, and c is equal to 1800 kg/m.

4.35 A: A) Start by writing a Newton's 2nd Law Equation in the horizontal direction, and then applying separation of variables and integrating both sides.

$$F_{net_x} = ma_x \rightarrow -cv^2 = m\frac{dv}{dt} \rightarrow \frac{dv}{v^2} = -\frac{c}{m}dt \rightarrow \int_{v_0}^v \frac{dv}{v^2} = \int_0^t -\frac{c}{m}dt$$

Now perform the integration and solve for the velocity.

$$\int_{v_0}^v \frac{dv}{v^2} = \int_0^t -\frac{c}{m}dt \rightarrow -\frac{1}{v}\Big|_{v_0}^v = -\frac{c}{m}t \rightarrow -\frac{1}{v} + \frac{1}{v_0} = -\frac{c}{m}t \rightarrow \frac{1}{v} = \frac{1}{v_0} + \frac{c}{m}t \rightarrow$$

$$\frac{1}{v} = \frac{m}{mv_0} + \frac{cv_0 t}{mv_0} = \frac{m + cv_0 t}{mv_0} \rightarrow v(t) = \frac{mv_0}{m + cv_0 t}$$

B) Substitute in your givens to find the speed at t=120 seconds.

$$v(t) = \frac{mv_0}{m + cv_0 t} \xrightarrow{\substack{m=10^7 kg \\ c=1800\frac{kg}{m} \\ v_0=20\frac{m}{s} \\ t=120s}} v(t) = \frac{(10^7 kg)(20\frac{m}{s})}{10^7 kg + (1800\frac{kg}{m})(20\frac{m}{s})(120s)} = 14\frac{m}{s}$$

Ramps and Inclines

Now that you've developed an understanding of Newton's Laws of Motion, free body diagrams, friction, and forces on flat surfaces, you can extend these tools to situations on ramps, or inclined surfaces. The key to understanding these situations is creating an accurate free body diagram after choosing convenient x- and y-axes. Problem-solving steps are consistent with those developed for Newton's 2nd Law.

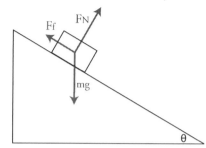

Take the example of a box on a ramp inclined at an angle of θ with respect to the horizontal. You can draw a basic free body diagram for this situation, with the force of gravity pulling the box straight down, the normal force perpendicular out of the ramp, and friction opposing motion (in this case pointing up the ramp).

Once the forces acting on the box have been identified, you must be clever about our choice of x-axis and y-axis directions. Much like analyzing free falling objects and projectiles, if you set the positive x-axis in the direction of initial motion (or the direction the object wants to move if it is not currently moving), the y-axis must lie perpendicular to the ramp's surface (parallel to the normal force). Now, you can re-draw the free body diagram, this time superimposing it on the new axes. You may also want to rotate the book slightly counter-clockwise until the x- and y-axes are horizontal and vertical again.

Unfortunately, the force of gravity on the box, mg, doesn't lie along one of the axes. Therefore, it must be broken up into components which do lie along the x- and y-axes in order to simplify the mathematical analysis. To do this, you can break the force of gravity vector down into a component parallel with the axis of motion ($mg_{\parallel}$) and a component perpendicular to the x-axis ($mg_{\perp}$) using trigonometry:

$$mg_{\parallel} = mg \sin \theta$$
$$mg_{\perp} = mg \cos \theta$$

You can now re-draw the free body diagram, replacing mg with its components. All the forces line up with the axes, making it straightforward to write Newton's 2nd Law Equations in the x- and y-directions and continue with your standard problem-solving strategy.

In the example shown with the modified free body diagram, you could write the Newton's 2nd Law Equations for both the x- and y-directions as follows:

$$F_{net_x} = mg_{\parallel} - F_f = mg \sin \theta - F_f = ma_x$$
$$F_{net_y} = F_N - mg_{\perp} = F_N - mg \cos \theta = 0$$

From this point, the problem becomes an exercise in algebra. If you need to tie the two equations together to eliminate a variable, don't forget the equation for the force of friction.

4.36 Q: A 5-kg mass is held at rest on a frictionless 30° incline by force F. What is the magnitude of F?

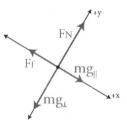

4.36 A: Start by identifying the forces on the box (below left) and making a free body diagram (below center).

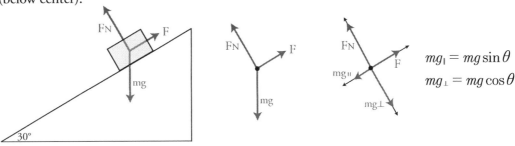

Break up the weight of the box into components parallel to and perpendicular to the ramp, and re-draw the free body diagram using the components of the box's weight (above right). Finally, use Newton's 2nd Law in the x-direction to solve for the force F.

$$F_{net_x} = ma_x \xrightarrow[\substack{F_{net_x}=F-mg_\parallel \\ mg_\parallel = mg\sin\theta \\ a_x=0}]{} F - mg\sin\theta = 0 \rightarrow F = mg\sin\theta = (5kg)(10^m\!/_{s^2})\sin(30°) = 25N$$

4.37 Q: A 10-kg box slides down a frictionless 18° ramp. Find the acceleration of the box, and the time it takes the box to slide 2 meters down the ramp.

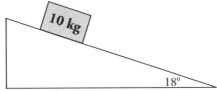

4.37 A: Start by identifying the forces on the box (below left) and making a free body diagram (below center).

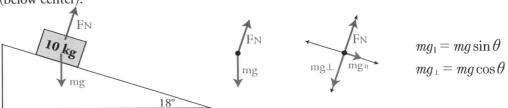

Break up the weight of the box into components parallel to and perpendicular to the ramp (above right). Then, use Newton's 2nd Law to find the acceleration.

$$F_{net_x} = ma_x \xrightarrow[\substack{F_{net_x}=mg_\parallel \\ mg_\parallel = mg\sin\theta}]{} mg\sin\theta = ma_x \rightarrow a_x = g\sin\theta = (10^m\!/_{s^2})\sin 18° = 3.09^m\!/_{s^2}$$

Finally, use the acceleration to solve for the time it takes the box to travel 2m down the ramp using kinematic equations.

$$\Delta x = v_{0_x}t + \tfrac{1}{2}a_x t^2 \xrightarrow[\substack{v_{0_x}=0}]{} t = \sqrt{\frac{2\Delta x}{a}} = \sqrt{\frac{2(2m)}{3.09^m\!/_{s^2}}} = 1.14s$$

Note that the solution has no dependence on the mass of the box.

4.38 Q: A block weighing 10 newtons is on a ramp inclined at 30° to the horizontal. A 3-newton force of friction, F_f, acts on the block as it is pulled up the ramp at constant velocity with force F, which is parallel to the ramp, as shown in the diagram. What is the magnitude of force F?

v (constant)

F

10.0 N

F_f=3.0N

Vectors are not drawn to scale.

30.0°

4.38 A: Start by identifying the forces on the box (below left) and making a free body diagram (below center). Then break up the weight of the box into components parallel to and perpendicular to the ramp (below right).

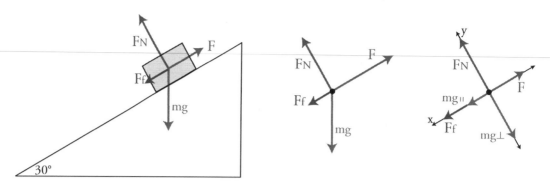

Apply Newton's second law, realizing that the forces must be balanced (net force is zero) since the box is moving at constant velocity.

$$F_{net_x} = ma_x \xrightarrow{a_x=0} F - F_f - mg_{\shortparallel} = 0 \xrightarrow[\substack{F_f=3N}]{mg_{\shortparallel}=mg\sin\theta} F = 3N + (10N)\sin(30°) = 8N$$

4.39 Q: A block of mass m is pulled up a ramp at constant speed by an applied force F as shown in the diagram below.

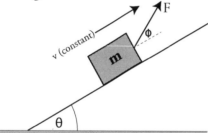

Which of the following is a possible expression for the coefficient of friction between the block and the ramp?

(A) $g\dfrac{\tan\theta}{\tan\phi}$

(B) $\dfrac{F\sin\theta}{mg\cos\theta - F\cos\phi}$

(C) $\dfrac{mg\cos\phi - F\sin\phi}{mg\cos\theta + F\cos\theta}$

(D) $\dfrac{mg\cos\theta - F\sin\phi}{mg\sin\theta + F\cos\phi}$

(E) $\dfrac{F\cos\phi - mg\sin\theta}{mg\cos\theta - F\sin\phi}$

Chapter 4: Dynamics

4.39 A: (E) Start by identifying the forces on the box (below left) and making a free body diagram (below center). Then break up the weight of the box into components parallel to and perpendicular to the ramp, and break up vector $\vec{F}$ into components parallel to and perpendicular to the ramp (below right).

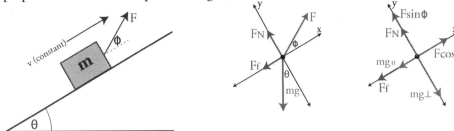

Next, write Newton's 2nd Law equations in the x- and y-directions to solve for the frictional force and the normal force.

$$F_{net_x} = ma_x \xrightarrow{a_x=0} F_f + mg\sin\theta - F\cos\phi = 0 \rightarrow F_f = F\cos\phi - mg\sin\theta$$

$$F_{net_y} = ma_y \xrightarrow{a_y=0} F\sin\phi + F_N - mg\cos\theta = 0 \rightarrow F_N = mg\cos\theta - F\sin\phi$$

Finally, solve for the coefficient of friction.

$$|\vec{F}_f| = \mu|\vec{F}_N| \rightarrow \mu = \frac{F_f}{F_N} = \frac{F\cos\phi - mg\sin\theta}{mg\cos\theta - F\sin\phi}$$

4.40 Q: A block of mass m is pushed up an incline with an initial speed of v_0. The incline has a height of h and the length of the ramp is L. The coefficient of kinetic friction between the incline and block is μ_k and the coefficient of static friction is μ_s. The block does not leave the ramp.

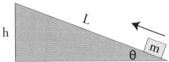

Use the variables listed above and fundamental constants to answer:
A) What is the angle of the ramp in terms of h and L?
B) What is the minimum angle of the ramp such that the block slides back down the ramp after sliding up? (Hint: the answer is not in terms of h and L.)
C) Assume the distance the block travels down the ramp is the same distance the block traveled up the ramp. Does it take more time for the block to travel up or down the ramp? Explain your answer.

4.40 A: A) $\theta = \sin^{-1}\left(\frac{h}{L}\right)$

B) When the block slows to a stop, the component of the weight along the incline has to be slightly larger than the force of friction in order to start back down the incline; therefore $F_f < mg\sin\theta$. Then, utilizing the equation for the force of friction and the static coefficient of friction since the block stops at the top of its path, $F_f = \mu_s F_N = \mu_s mg\cos\theta$. Then, set the two force of friction equations equal to each other and solve for the angle θ: $\mu_s mg\cos\theta = mg\sin\theta \rightarrow \theta = \tan^{-1}(\mu_s)$.

C) It takes more time for the block to travel down the ramp since it is moving at a slower speed on the way down due to the force of friction acting as a nonconservative force. This work the friction does on the block decreases the kinetic energy of the block, so the longer the block slides on the ramp, the more energy that is being removed from the ramp block system.

Newton's 3rd Law of Motion

Newton's 3rd Law of Motion, commonly referred to as the Law of Action and Reaction, describes the phenomena by which all forces come in pairs. A force is really an interaction between two objects. If Object 1 exerts a force on Object 2, then Object 2 must exert a force back on Object 1, and the force of Object 1 on Object 2 is equal in magnitude, or size, but opposite in direction to the force of Object 2 on Object 1. Written mathematically:

$$\vec{F}_{1\,on\,2} = -\vec{F}_{2\,on\,1}$$

This has many implications, some of which aren't immediately obvious. For example, if you punch the wall with your fist with a force of 100N, the wall imparts a force back on your fist of 100N (which is why it hurts!). Or try this: push on the corner of your desk with your palm for a few seconds. Now look at your palm… see the indentation? That's because the corner of the desk pushed back on your palm.

Although this law surrounds your actions everyday, oftentimes you may not even realize its effects. To run forward, a cat pushes with its legs backward on the ground, and the ground pushes the cat forward. How do you swim? If you want to swim forwards, which way do you push on the water? Backwards, that's right. As you push backwards on the water, the reactionary force, the water pushing you, propels you forward. How do you jump up in the air? You push down on the ground, and it's the reactionary force of the ground pushing on you that accelerates you skyward!

As you can see, then, forces always come in pairs because forces result from the interaction of two objects. These pairs are known as **action-reaction pairs**. With this in mind, you can see why an object can never exert a force upon itself; there must be another object to interact with in order to generate force pairs. What are the action-reaction force pairs for a girl kicking a soccer ball? The girl's foot applies a force on the ball, and the ball applies an equal and opposite force on the girl's foot. Further, if two objects interact with each other by applying a force upon each other, and the objects are part of the same system, the velocity of the system's center of mass cannot change.

How does a rocket ship maneuver in space? The rocket propels hot expanding gas particles outward, so the gas particles in return push the rocket forward. Newton's 3rd Law even applies to gravity. The Earth exerts a gravitational force on you (downward). You, therefore, must apply a gravitational force upward on the Earth!

4.41 Q: Earth's mass is approximately 81 times the mass of the Moon. If Earth exerts a gravitational force of magnitude F on the Moon, the magnitude of the gravitational force of the Moon on Earth is
(A) F
(B) F/9
(C) F/81
(D) 9F
(E) 81F

4.41 A: (A) The force Earth exerts on the Moon is the same in magnitude and opposite in direction of the force the Moon exerts on Earth.

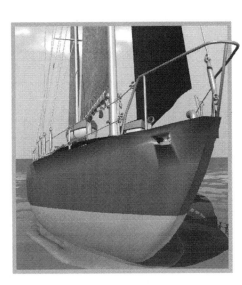

4.42 Q: A 400-newton girl standing on a dock exerts a force of 100 newtons on a 10,000-newton sailboat as she pushes it away from the dock. How much force does the sailboat exert on the girl?
(A) 0.25 N
(B) 25 N
(C) 100 N
(D) 400 N
(E) 10,000 N

4.42 A: (C) The force the girl exerts on the sailboat is the same in magnitude and opposite in direction of the force the sailboat exerts on the girl.

4.43 Q: A carpenter hits a nail with a hammer. Compared to the magnitude of the force the hammer exerts on the nail, the magnitude of the force the nail exerts on the hammer during contact is
(A) less
(B) greater
(C) the same

4.43 A: (C) the same per Newton's 3rd Law.

4.44 Q: Identical fireflies are placed in closed jars in three different configurations as shown below. In configuration A, three fireflies are hovering inside the jar. In configuration B, one firefly is hovering inside the jar. In configuration C, one firefly is sitting at rest on the bottom of the jar. Each jar is placed upon a scale and measured. Rank the weight of each jar according to the scale reading from heaviest to lightest. If jars have the same scale reading, rank them equally.

| A | B | C |

4.44 A: A, B=C. Whether the fireflies are in flight or sitting on the bottom of the jar, they provide the same weight to the scale (if they are flying in the jar, the force their wings provide on the air pushing down is equal to the force of the air pushing them up. This same air pushes down on the bottom of the jar by Newton's 3rd Law, making their weights equivalent whether flying or resting. Therefore, the only factor in determining the weight is the number of fireflies in the jar.

4.45 Q: A physics teacher poses this question to her class: "If forces only come in pairs that are equal and opposite, why don't all forces cancel each other so that there's no acceleration?" Which, if any, of the following student answers do you agree with?

- Alfred: *"In specific situations, Newton's 2nd Law supercedes Newton's 3rd Law, so that an unbalanced force on an object can create an acceleration."*

- Bethany: *"All forces are equal and opposite, but they act on different objects, so individual objects can have unbalanced forces that create an acceleration."*

- Charlie: *"All forces do come in pairs and cancel each other out, but the acceleration of an object also acts like an additional force."*

4.45 A: Bethany is correct. Forces always come in pairs that are equal and opposite, but these forces act on separate objects.

4.46 Q: Paisley the horse gets stuck in the mud. Her rider, Linda, ties a rope around Paisley and pulls with a force of 500 newtons, but isn't strong enough to get Paisley out of the mud. In a flash of inspiration, Linda thinks back to her physics classes and ties one end of the rope to Paisley, and the other end to a nearby fence post. She then applies the same force of 500 newtons to the middle of the rope at an angle of $\theta = 6°$, just barely freeing Paisley from the mud, as shown in the diagram below.

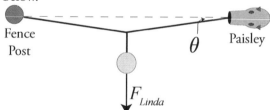

Determine the minimum amount of tension the rope must support.

4.46 A: A free body diagram showing all the forces on the rope is quite helpful in this situation.

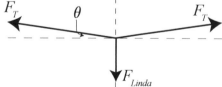

Note that the tension in the portion of rope connected to the fence post must be the same as the tension in the portion of rope connected to Paisley.

$$F_{net_y} = 2F_T \sin\theta - F_{Linda} = 0 \rightarrow F_T = \frac{F_{Linda}}{2\sin\theta} = \frac{500N}{2\sin 6°} = 2400N$$

Pulleys and Atwood Machines

Ideal pulleys allow you to redirect forces without adding friction or inertia to the system under observation. Later, in the rotation chapter, you'll study non-ideal pulleys. **Atwood Machines** are basic physics laboratory devices often used to demonstrate basic principles of dynamics and acceleration. The machines typically involve a pulley, a string, and a system of masses. Keys to solving Atwood Machine problems are recognizing that the force transmitted by a string or rope, known as tension, is constant throughout the string, and choosing a consistent direction as positive. Let's walk through an example to demonstrate.

4.47 Q: Two masses, m_1 and m_2, are hanging by a massless string from a frictionless pulley. If m_1 is greater than m_2, determine the acceleration of the two masses when released from rest.

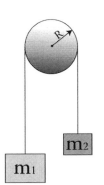

4.47 A: First, identify a direction as positive. Since you can easily observe that m_1 will accelerate downward and m_2 will accelerate upward, since $m_1 > m_2$, call the direction of motion around the pulley and down toward m_1 the positive y direction. Then, you can create free body diagrams for both object m_1 and m_2, as shown below:

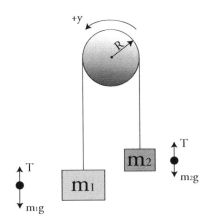

Using this diagram, write Newton's 2nd Law equations for both objects, taking care to note the positive y direction:

(1) $m_1g - T = m_1a$

(2) $T - m_2g = m_2a$

Next, combine the equations and eliminate T by solving for T in equation (2) and substituting in for T in equation (1).

(2) $T - m_2g = m_2a \rightarrow T = m_2g + m_2a$

(1) $m_1g - T = m_1a \xrightarrow{T=m_2g+m_2a}$

$m_1g - m_2g - m_2a = m_1a$

Finally, solve the resulting equation for the acceleration of the system.

$m_1g - m_2g - m_2a = m_1a \rightarrow m_1g - m_2g = m_1a + m_2a \rightarrow g(m_1 - m_2) = a(m_1 + m_2) \rightarrow$

$a = g\dfrac{(m_1 - m_2)}{(m_1 + m_2)}$

Alternately, you could treat both masses as part of the same system.

Drawing a dashed line around the system, you can directly write an appropriate Newton's 2nd Law equation for the entire system.

$m_1g - m_2g = (m_1 + m_2)a \rightarrow a = g\dfrac{(m_1 - m_2)}{(m_1 + m_2)}$

Note that if the string and pulley were not massless, this problem would get considerably more involved.

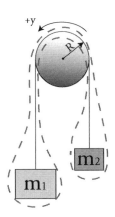

4.48 Q: Two masses, m_1 and m_2, are connected by a light string over a massless pulley of radius R as shown. Assuming a frictionless surface, find the acceleration of m_2.

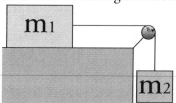

4.48 A: First, draw a free body diagram for each of the two masses. Call the clockwise direction around the pulley the positive y direction.

Next, write Newton's 2nd Law equations for each mass and solve for the acceleration of m_2, recognizing that the magnitude of the horizontal acceleration of m_1 must equal the magnitude of the vertical acceleration of m_2.

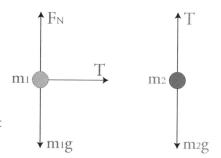

Because the pulley is massless, you can state that the tension must be uniform everywhere in the light string.

$$F_{net_x} = ma_x \rightarrow T = m_1 a$$

$$F_{net_y} = ma_y \rightarrow m_2 g - T = m_2 a \xrightarrow{T = m_1 a} m_2 g - m_1 a = m_2 a \rightarrow a = g\left(\frac{m_2}{m_1 + m_2}\right)$$

4.49 Q: Two masses are hung from a light string over an ideal frictionless massless pulley. The masses are shown in various scenarios in the diagram below. Rank the acceleration of the masses from greatest to least.

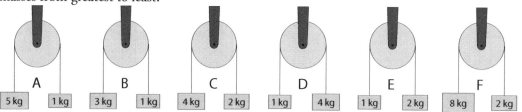

4.49 A: A > D=F > B > C=E

A simple analysis of the Atwood Machines shows that the acceleration of the system is equal to the net force divided by the sum of the masses. Summarizing in table form:

Scenario	m_1 (kg)	m_2 (kg)	F_{net} (N)	a (m/s²)
A	5	1	40	6.67
B	3	1	20	5
C	4	2	20	3.33
D	1	4	30	6
E	1	2	10	3.33
F	8	2	60	6

Centripetal Force

If an object traveling in a circular path has an inward acceleration, Newton's 2nd Law states there must be a net force directed toward the center of the circle as well. This type of force, known as a **centripetal force**, can be a gravitational force, a tension, an applied force, or even a frictional force.

> **Hint:** When dealing with circular motion problems, it is important to realize that a centripetal force isn't really a new force. A centripetal force is just a label or grouping you apply to a force to indicate its direction is toward the center of a circle. This means that you never want to label a force on a free body diagram as a centripetal force, $\vec{F}_c$. Instead, label the center-directed force as specifically as you can. If a tension is causing the force, label the force $\vec{T}$ or $\vec{F}_T$. If a frictional force is causing the center-directed force, label it $\vec{F}_f$, and so forth. Because a centripetal force is always perpendicular to the object's motion, a centripetal force can do no work on an object.

You can combine the equation for centripetal acceleration with Newton's 2nd Law to obtain Newton's 2nd Law for Circular Motion. Recall that Newton's 2nd Law states:

$$\vec{F}_{net} = m\vec{a}$$

For an object traveling in a circular path, there must be a net (centripetal) force directed toward the center of the circular path to cause a (centripetal) acceleration directed toward the center of the circular path. You can revise Newton's 2nd Law for this particular case as follows:

$$\vec{F}_{net_c} = m\vec{a}_c$$

Then, recalling the formula for the magnitude of the centripetal acceleration as:

$$a_c = \frac{v^2}{r}$$

You can put these together, replacing a_c in the equation to get a combined form of Newton's 2nd Law for Uniform Circular Motion:

$$|\vec{F}_{net_c}| = \frac{m|\vec{v}|^2}{r} \quad or \quad F_{net_c} = \frac{mv^2}{r}$$

Of course, if an object is traveling in a circular path and the centripetal force is removed, the object will continue traveling in a straight line in whatever direction it was moving at the instant the force was removed.

4.50 Q: An 800N star running back turns a corner in a circular path of radius 1 meter at a speed of 8 m/s. Find the running back's mass, centripetal acceleration, and centripetal force.

4.50 A: $mg = 800N \xrightarrow{g=10^m/_{s^2}} m = \dfrac{800N}{10^m/_{s^2}} = 80kg$

$a_c = \dfrac{v^2}{r} = \dfrac{(8^m/_s)^2}{1m} = 64^m/_{s^2}$

$F_{net_c} = ma_c = (80kg)(64^m/_{s^2}) = 5100N$

Note that you have found the magnitude of the centripetal acceleration and centripetal force vectors. The direction of the centripetal acceleration and centripetal force vectors must be toward the center of the running back's circular path.

4.51 Q: A 65-kg rider on an amusement park known as "The Rotor" spins inside a giant cylinder of radius 2.5 meters at a constant speed of 8.6 m/s. The floor is lowered and the student remains against the wall without falling to the floor.
A) Determine the magnitude of the centripetal force on the rider and describe the source of the centripetal force.
B) What upward force keeps the rider from falling when the floor is lowered?
C) What would happen to the frictional force on the rider if the radius of the ride and the speed of the ride were both halved?

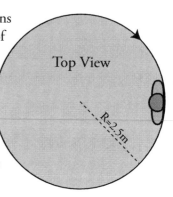

Top View

R=2.5m

4.51 A: A) The normal force of the wall on the rider provides the centripetal force.

$$F_{net_c} = F_N = \frac{mv^2}{r} = \frac{(65kg)(8.6^{m}/_{s^2})^2}{(2.5m)} = 1920N$$

B) The force of friction acts upward, balancing the gravitational force down on the rider.
C) If the radius of the ride and the speed of the ride were both halved, the centripetal force, caused by the normal force, would be halved. Since the frictional force is directly proportional to the normal force ($|\vec{F}_f| = \mu|\vec{F}_N|$), the frictional force would also be halved.

4.52 Q: Curves can be banked at just the right angle that vehicles traveling a specific speed can stay on the road with no friction required. Given a radius for the curve and a specific velocity, at what angle should the bank of the curve be built?

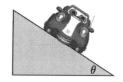

4.52 A: First draw a free body diagram for the car on the banked curve, then a pseudo-FBD, showing all forces lining up parallel to the axes. In this case, you do NOT tilt the x- and y-axes, since the direction the car is moving isn't up or down the ramp, but rather toward the center of the curve (in the positive x-direction when the car is in the position shown in the bottom left diagram). It is the normal force that doesn't line up with the axes, which is then broken up into components along the x- and y-axes in the pseudo-FBD.

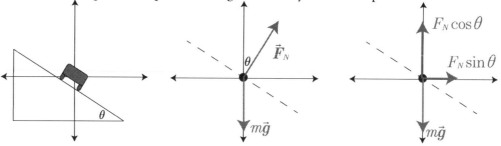

Then, write out Newton's 2nd Law equations for both the x- and y-directions.

$$F_{net_x} = F_{net_c} = F_N \sin\theta = ma_c = \frac{mv^2}{r}$$

$$F_{net_y} = F_N \cos\theta - mg = ma_y = 0 \rightarrow F_N = \frac{mg}{\cos\theta}$$

Finally, combining the results of these two equations, solve for the angle θ.

$$F_N \sin\theta = \frac{mv^2}{r} \xrightarrow{F_N = \frac{mg}{\cos\theta}} \frac{mg\sin\theta}{\cos\theta} = \frac{mv^2}{r} \xrightarrow{\frac{\sin\theta}{\cos\theta} = \tan\theta} g\tan\theta = \frac{v^2}{r} \rightarrow \theta = \tan^{-1}\left(\frac{v^2}{gr}\right)$$

Vertical Circular Motion

Objects travel in circles vertically as well as horizontally. Because the speed of these objects isn't typically constant, technically this isn't uniform circular motion, but your force and UCM analysis skills still prove applicable.

Consider a roller coaster traveling in a vertical loop of radius 10m. You travel through the loop upside down, yet you don't fall out of the roller coaster. How is this possible? You can use your understanding of UCM and dynamics to find out!

To begin with, first take a look at the coaster when the car is at the bottom of the loop. Drawing a free body diagram, the force of gravity on the coaster, also known as its weight, pulls it down. Opposing that force is the normal force of the rails of the coaster pushing up.

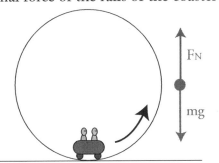

Because the coaster is moving in a circular path, you can analyze it using the tools developed for uniform circular motion. Newton's 2nd Law still applies, so you can write:
$$F_{net_c} = F_N - mg$$

Notice that because you're talking about circular motion, you can adopt the convention that forces pointing toward the center of the circle are positive, and forces pointing away from the center of the circle are negative. At this point, recall that the force you "feel" when you're in motion is actually the normal force. So, solving for the normal force as you begin to move in a circle, you find:
$$F_N = F_{net_c} + mg$$

Since you know that the net force is always equal to mass times acceleration, the net centripetal force is equal to mass times the centripetal acceleration. You can therefore write:
$$F_N = F_{net_c} + mg = \frac{mv^2}{r} + mg$$

You can see from the resulting equation that the normal force is now equal to the weight plus an additional term from the centripetal force of the circular motion. As you travel in a circular path near the bottom of the loop, you feel heavier than your weight. In common terms, you feel additional "g-forces." How many g's you feel can be obtained with a little bit more manipulation. If you re-write your equation for the normal force, pulling out the mass by applying the distributive property of multiplication, you obtain:
$$F_N = m\left(\frac{v^2}{r} + g\right)$$

Notice that inside the parenthesis you have the standard acceleration due to gravity, g, plus a term from the centripetal acceleration. This additional term is the additional g-force felt by a person. For example, if a_c was equal to g (~10 m/s²), you could say the person in the cart was experiencing two g's (1g from the centripetal acceleration, and 1g from the Earth's gravitational field). If a_c were equal to 3×g (~30 m/s²), the person would be experiencing a total of four g's.

Expanding this analysis to a similar situation in a different context, try to imagine instead of a roller coaster, a mass whirling in a vertical circle by a string. You could replace the normal force by the tension in the string in the analysis. Because the force is larger at the bottom of the circle, the likelihood of the string breaking is highest when the mass is at the bottom of the circle!

At the top of the loop, you have a considerably different picture. Now, the normal force from the coaster rails must be pushing down against the cart, though still in the positive direction since down is now toward the center of the circular path. In this case, however, the weight of the object also points toward the center of the circle, since the Earth's gravitational field always pulls toward the center of the Earth. The FBD looks considerably different, and therefore the application of Newton's 2nd Law for Circular Motion is considerably different as well:

$$F_{net_c} = F_N + mg$$

Since the force you feel is actually the normal force, you can solve for the normal force and expand the net centripetal force as shown:

$$F_N = F_{net_c} - mg = \frac{mv^2}{r} - mg$$

You can see from the equation that the normal force is now the centripetal force minus your weight. If the centripetal force were equal to your weight, you would feel as though you were weightless. Note that this is also the point where the normal force is exactly equal to 0. This means the rails of the track are no longer pushing on the roller coaster cart. If the centripetal force was slightly smaller, and the cart's speed was slightly smaller, the normal force F_N would be less than 0. Since the rails can't physically pull the cart in the negative direction (away from the center of the circle), this means the car is falling off the rail and the cart's occupant is about to have a very, very bad day. Only by maintaining a high speed can the cart successfully negotiate the loop. Go too slow and the cart falls.

In order to remain safe, real roller coasters actually have wheels on both sides of the rails to prevent the cart from falling if it ever did slow down at the top of a loop, although coasters are designed so that this situation never actually occurs.

Photo © Ruslan Kerimov | Dreamstime.com

4.53 Q: A sphere is attached to the end of a string of length R as shown in the diagram.

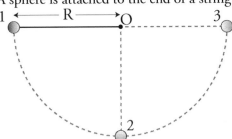

The end of the string is attached at origin point O. The sphere is released from rest at position 1, swinging through position 2 and up to position 3. What is the direction of the net force on the sphere at each of the three positions?

4.53 A: At position 1, the sphere is at rest and experiences a downward force of gravity. At point 2, the sphere is traveling in a circular path and experiences a net force toward point O at the center of the circle. At point 3, the sphere again comes to rest, experiencing a net force down due to gravity.

4.54 Q: A roller coaster has a loop of radius 6 m. The roller coaster cars have an average mass of 5000 kg with riders, and travel at a speed of 15 m/s at the top of the loop, and a speed of 20 m/s at the bottom of the loop.
A) Determine the amount of force the track must be designed to withstand at the top in order to keep the cars going around the loop.
B) Determine the amount of force the track must be designed to withstand at the bottom of the loop.
C) Determine the minimum speed the cars on the roller coaster can move and still make it through the loop without losing contact with the rails.

4.54 A: A) At the highest point, the free body diagram has both the weight of the cars and the normal force pointing toward the center of the circle, as shown on the FBDs below.

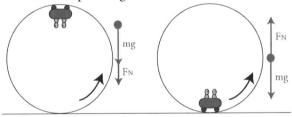

$$F_{net_c} = F_N + mg = ma_c \rightarrow F_N = ma_c - mg \xrightarrow{a_c = \frac{mv^2}{r}} F_N = \frac{mv^2}{r} - mg \rightarrow$$

$$F_N = \frac{(5000kg)(15\,{}^m/_s)^2}{(6m)} - (5000kg)(10\,{}^m/_{s^2}) = 138{,}000N$$

B) At the lowest point, the weight still pulls down, but the normal force pushes up:

$$F_{net_c} = F_N - mg = ma_c \rightarrow F_N = ma_c + mg \xrightarrow{a_c = \frac{mv^2}{r}} F_N = \frac{mv^2}{r} + mg \rightarrow$$

$$F_N = \frac{(5000kg)(15\,{}^m/_s)^2}{(6m)} + (5000kg)(10\,{}^m/_{s^2}) = 383{,}000N$$

C) The minimum speed the cars on the roller coast can move occurs when the normal force is just barely equal to 0N when the car is at the top of the loop. Use this to solve for the velocity.

$$F_N = 0 \xrightarrow{top\ position} F_N = ma_c - mg = 0 \rightarrow ma_c = mg \rightarrow \frac{v^2}{r} = g \rightarrow v = \sqrt{gr} \rightarrow$$

$$v = \sqrt{(10\,{}^m/_{s^2})(6m)} = 7.75\,{}^m/_s$$

Chapter 4: Dynamics

Chapter 5: Work, Energy, and Power

Declaration, July 4th 1776

"Energy and persistence conquer all things."

—Benjamin Franklin

Objectives

1. Calculate the work done by a force on an object as it is displaced.

2. Determine the work done from a graph of force vs. displacement.

3. Calculate the change in the kinetic energy of an object as a result of work being done on or by that object.

4. Explain what it means for a force to be conservative.

5. Utilize the relationship between conservative forces and potential energy to solve problems.

6. Apply Hooke's Law to determine the elastic potential energy stored in a spring.

7. Determine the gravitational potential energy of a system in a gravitational field.

8. Apply the law of conservation of energy to a variety of situations involving both conservative and non-conservative forces.

9. Solve problems involving both the law of conservation of energy and Newton's Laws of Motion.

10. Determine the power required and/or dissipated to accomplish specific tasks.

W̱ork, energy and power are highly inter-related concepts that come up regularly in everyday life. You do work on an object when you move it. The rate at which you do the work is your power output. When you do work on an object, you transfer energy from one object to another. In this chapter you'll explore how energy is transferred and transformed, how doing work on an object changes its energy, and how quickly work can be done.

Work

In physics terms, **work** is the process of moving an object by applying a force, or, more formally, work is the energy transferred by an external force exerted on an object or system that moves the object or system. Work is a scalar quantity—it has magnitude only, not direction. In order for work to be done, the object must move, and the force must cause the movement.

The units of work can be found by performing unit analysis on the work formula. If work is force multiplied by distance, the units must be the units of force multiplied by the units of distance, or newtons multiplied by meters. A newton-meter is also known as a Joule (J).

$$1 J = 1 N \cdot m = 1 \frac{kg \cdot m^2}{s^2}$$

I'm sure you can think up countless examples of work being done, but a few that spring to mind include pushing a snowblower to clear the driveway, pulling a sled up a hill with a rope, stacking boxes of books from the floor onto a shelf, and throwing a baseball from the pitcher's mound to home plate.

Let's take a look at a few scenarios and investigate what work is being done.

In the first scenario, a monkey in a jet pack blasts through the atmosphere, accelerating to higher and higher speeds. In this case, the jet pack is applying a force causing it to move. But what is doing the work? Hot expanding gases are pushed backward out of the jet pack. Using Newton's 3rd Law, you observe the reactionary force of the gas pushing the jet pack forward, causing a displacement. Therefore, the expanding exhaust gas is doing work on the jet pack.

In the second scenario, a girl struggles to push her stalled car, but can't make it move. Even though she's expending significant effort, no work is being done on the car because it isn't moving.

In the final scenario, a child in a ghost costume carries a bag of Halloween candy across the yard. In this situation, the child applies a force upward on the bag, but the bag moves horizontally. From this perspective, the forces of the child's arms on the bag don't cause the displacement; therefore, no work is done by the child.

It's important to note that when calculating work, only the force applied in the direction of the object's displacement counts! This means that if the force and displacement vectors aren't in exactly the same direction, you need to take the component of force in the direction of the object's displacement.

For the case of a constant force, this can be accomplished by taking the dot product of the force and displacement vectors. Take the example of a box pushed along a floor through a displacement $\Delta \vec{x}$ by application of force $\vec{F}$. You can find the work done by the force as follows:

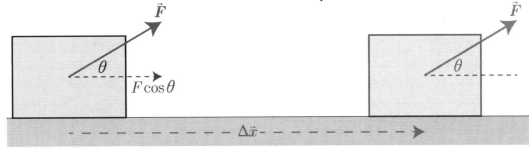

$$W = \vec{F} \cdot \Delta \vec{x} = |\vec{F}||\Delta \vec{x}|\cos\theta$$

5.01 Q: An appliance salesman pushes a refrigerator 2 meters across the floor by applying a force of 200N. Find the work done.

5.01 A: Since the force and displacement are in the same direction, the angle between them is 0.
$$W = \vec{F} \cdot \Delta \vec{x} = |\vec{F}||\Delta \vec{x}|\cos\theta = (200N)(2m)\cos(0°) = 400J$$

5.02 Q: A friend's car is stuck on the ice. You push down on the car to provide more friction for the tires (by way of increasing the normal force), allowing the car's tires to propel it forward 5m onto less slippery ground. How much work did you do?

5.02 A: You applied a downward force, yet the car's displacement was sideways. Therefore, the angle between the force and displacement vectors is 90°.
$$W = \vec{F} \cdot \Delta \vec{x} = |\vec{F}||\Delta \vec{x}|\cos\theta = |\vec{F}||\Delta \vec{x}|\cos(90°) = 0J$$

5.03 Q: How much work is done in lifting an 8-kg box from the floor to a height of 2m above the floor?

5.03 A: It's easy to see the displacement is 2m, and the force must be applied in the direction of the displacement, but what is the force? To lift the box you must match and overcome the force of gravity on the box. Therefore, the force applied is equal to the gravitational force, or weight, of the box, mg=(8kg)(10m/s²)=80N.
$$W = \vec{F} \cdot \Delta \vec{x} = |\vec{F}||\Delta \vec{x}|\cos\theta = (80N)(2m)\cos(0°) = 160J$$

5.04 Q: You push a crate up a ramp with a force of 10N. Despite your pushing, however, the crate slides down the ramp a distance of 4m. How much work did you do?

5.04 A: Since the direction of the force you applied is opposite the direction of the crate's displacement, the angle between the two vectors is 180°.
$$W = \vec{F} \cdot \Delta \vec{x} = |\vec{F}||\Delta \vec{x}|\cos\theta = (10N)(4m)\cos(180°) = -40J$$

Four carts, initially at rest on a flat surface, are subjected to varying forces as the carts move to the right a set distance, depicted in the diagram below.

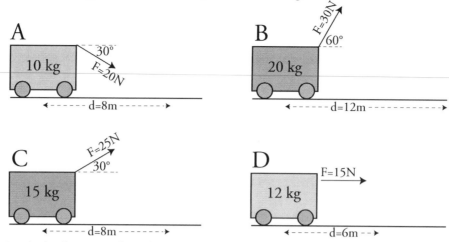

Rank the four carts from least to greatest in terms of
I) work done by the applied force on the carts
II) inertia
III) normal force applied by the surface to the carts

5.05 A: This is a great problem to make a table of information to assist in your rankings.

Cart	$|\vec{F}|$	θ	$|\Delta\vec{x}|$	W	m	$|\vec{F}_N|$
Cart A	20 N	30°	8 m	139 J	10 kg	110 N
Cart B	30 N	60°	12 m	180 J	20 kg	174 N
Cart C	25 N	30°	8 m	173 J	15 kg	138 N
Cart D	15 N	0°	6 m	90 J	12 kg	120 N

I) D, A, C, B
II) A, D, C, B
III) A, D, C, B

Non-Constant Forces

The area under a force vs. displacement graph is the work done by the force. Consider the situation of a block being pulled across a table with a constant force of 5 newtons over a displacement of 5 meters, then the force gradually tapers off over the next 5 meters.

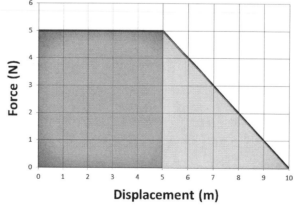

The work done by the force moving the block can be calculated by taking the area under the force vs. displacement graph (a combination of a rectangle and triangle) as follows:

$$W = A_{rectangle} + Area_{triangle} = lw + \tfrac{1}{2}bh = (5m)(5N) + \tfrac{1}{2}(5m)(5N) = 37.5J$$

This can easily be extended to deal with forces whose area isn't quite so easy to calculate. Given the force vs displacement graph below left, you can calculate the work done by the force on the object from initial point x_i to final point x_f by taking the area under the graph using your calculus skills (below center).

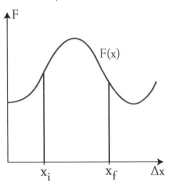 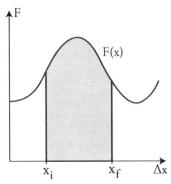 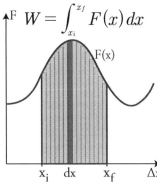

Break up the graph into infinitessimally small displacements along the x-axis, dx, and add up the areas of all infinitessimally thin rectangles to find the total area, which is the work done on the displaced object by the force (above right).

5.06 Q: The force on an object is given by the equation $F(x)=20x-2x^2$, where F is in newtons and x is in meters. Determine the work done on the object in moving from x=1 to x=3 meters.

5.06 A: $W = \int_{x_i}^{x_f} F(x)\,dx = \int_{1m}^{3m} (20x - 2x^2)\,dx = \left(20\tfrac{x^2}{2} - 2\tfrac{x^3}{3}\right)\Big|_1^3 = \left(10x^2 - \tfrac{2}{3}x^3\right)\Big|_1^3 \rightarrow$

$W = \left(10(3)^2 - \tfrac{2}{3}(3)^3\right) - \left(10(1)^2 - \tfrac{2}{3}(1)^3\right) = (90 - 18) - (10 - \tfrac{2}{3}) = 62.7J$

Hooke's Law

A common application of non-constant forces involves the modeling of springs. The more you stretch or compress a spring, the greater the force of the spring. The spring's force is opposite the direction of its displacement from equilibrium. Modeling this force as a linear relationship, where force applied by the spring is equal to some constant (known as the **spring constant**, k) multiplied by the spring's displacement from its equilibrium (rest) position, you arrive at what is commonly known as **Hooke's Law**:

$$\vec{F}_{spring} = -k\Delta\vec{x}$$

Although Hooke's Law isn't formally a law of physics, it is a reasonable model for many elastic forces at low levels of stress and strain.

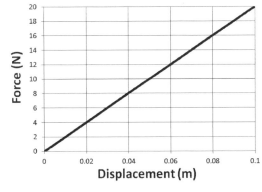

You can determine the spring constant of a spring by making a graph of the force from a spring on the y-axis, and placing the displacement of the spring from its equilibrium, or rest position, on the x-axis. The slope of the graph will give you the spring constant. For the case of the spring depicted in the graph at left, you can find the spring constant as follows:

$$k = slope = \frac{rise}{run} = \frac{\Delta F}{\Delta x} = \frac{20N - 0N}{0.1m - 0m} = 200\,{}^{N}\!/_{m}$$

You must have done work to compress or stretch the spring, since you applied a force and caused a displacement. You can find the work done in stretching or compressing a spring by taking the area under the graph. For the spring shown, to displace the spring 0.1m, you can find the work done as shown below:

$$W = A_{tri} = \tfrac{1}{2}bh = \tfrac{1}{2}(0.1m)(20N) = 1J$$

5.07 Q: A spring is subjected to a varying force and its elongation is measured and recorded in the table below. Determine the spring constant of the spring.

Force (N)	0	1.0	3.0	4.0	5.0	6.0
Elongation (m)	0	0.30	0.67	1.00	1.30	1.50

5.07 A: You can find the spring constant by plotting the data and determining the slope of the best-fit line.

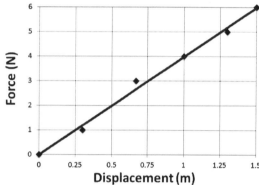

$$k = slope = \frac{rise}{run} = \frac{5N - 0N}{1.25m - 0m} = 4\,{}^{N}\!/_{m}$$

5.08 Q: The Atwood machine shown to the right consists of a massless, frictionless pulley, a massless string, and a massless spring. The spring has a spring constant of 200 N/m. How far will the spring stretch when the masses are released?

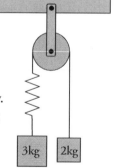

5.08 A: First solve for the acceleration of the masses by using Newton's second law. The 3-kg mass is accelerating downward (negative direction) and the 2-kg mass is accelerating upward (positive direction).

$$F_{net_y} = ma_y \rightarrow -30N + 20N = (5kg)\,a \rightarrow a = -2\,{}^{m}\!/_{s^2}$$

Next, isolate the 3-kg mass to find the tension acting on the spring. Again use Newton's second law.

$$F_{net_y} = ma_y \rightarrow F_T - 30N = (3kg)(-2\,{}^{m}\!/_{s^2}) \rightarrow F_T = 24N$$

Finally, use Hooke's Law to solve for the stretch of the spring.

$$|F_s| = kx \rightarrow 24N = (200\,{}^{N}\!/_{m})(x) \rightarrow x = 0.12m$$

5.09 Q: The Atwood machine shown to the right consists of a massless, frictionless pulley, a massless string, and a massless spring. The spring has a spring constant of 100 N/m. Calculate the stretch of the spring in terms of M.

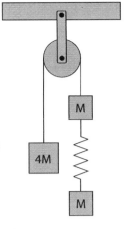

5.09 A: First solve for the acceleration of the masses using Newton's Second Law. The 4m mass is accelerating downward and the other 2 masses are accelerating upward. You can put these masses on the horizontal axis and assume the masses are all accelerating in the same –x direction.

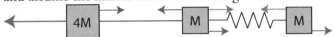

This new horizontal system can be analyzed using Newton's second law. Assume g = 10 N/kg. Analyzing forces from right to left:

$$F_{net} = ma \rightarrow -4Mg + F_T - F_T + Mg + F_s - F_s + Mg = (6M)\,a \rightarrow a = -\frac{g}{3} = -3.3\,{}^m\!/_{s^2}$$

Next, isolate the mass on the right end to find the force acting from the spring. Again use Newton's second law.

$$F_{net} = ma \rightarrow Mg - F_s = Ma \rightarrow F_s = Mg - Ma = (13.3M)$$

Finally, use Hooke's Law to solve for the stretch of the spring.

$$F_s = kx \rightarrow 13.3M = 100x \rightarrow x = 0.133M \;\; meters$$

Going further, assume a spring obeys Hooke's Law, where $|\vec{F}_{spring}(x)| = k|\vec{x}|$, or simply F(x)=kx. You can determine how much work is done in compressing or stretching the spring from its equilibrium position to some point x by integrating as follows:

$$W = \int \vec{F} \cdot d\vec{x} = \int_0^x kx\,dx = k \int_0^x x\,dx = k\frac{x^2}{2}\Big|_0^x = \tfrac{1}{2}kx^2$$

This implies that the work you do in compressing the spring some amount x is equal to half the spring constant multiplied by the square of the spring's displacement from equilibrium, and as you'll verify shortly, if you put this much work into the spring displacing it, the spring will have that same amount of elastic potential energy stored in it once it has been displaced that amount x.

5.10 Q: The force required to extend a non-linear spring is described by $F(x) = \tfrac{1}{2}kx^2$. How much work is done in compressing the spring from equilibrium to some point x?

5.10 A: $W = \int_0^x \vec{F} \cdot d\vec{x} = \int_0^x \tfrac{1}{2}kx^2\,dx = \frac{k}{2}\int_0^x x^2\,dx = \frac{k}{2}\frac{x^3}{3}\Big|_0^x = \frac{k}{2}\left(\frac{x^3}{3} - \frac{0^3}{3}\right) = \frac{kx^3}{6}$

Work in Multiple Dimensions

It is also possible (and quite common) to have situations in which an object is moving along a path in multiple dimensions, and a force applied to that object isn't necessarily in the same direction as the path. An illustration of just such a situation is shown below.

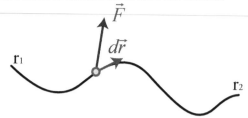

Assume the object moves along the given path from position r_1 to position r_2 (corresponding to position vectors $\vec{r}_1$ and $\vec{r}_2$). You can break up the path of the object into infinitessimally small pieces $d\vec{r}$, with a force $\vec{F}$ acting on the object at that portion of the path. The work done by the force in this infinitessimally small section of the path (dW) can be found from the initial equation for work.

$$dW = \vec{F} \cdot d\vec{r}$$

Finding the total work done in traveling from r_1 to r_2 can then be found by simply adding up all the little tiny bits of work (dW) along the path from r_1 to r_2, giving you the general formula for work.

$$W = \int_{\vec{r}_1}^{\vec{r}_2} dW \xrightarrow{dW = \vec{F} \cdot d\vec{r}} W = \int_{\vec{r}_1}^{\vec{r}_2} \vec{F} \cdot d\vec{r}$$

This type of integral is known as a line integral, as you are integrating, or adding up, all the little pieces of a function along a path. It's important to note that although it is called a line integral, the path doesn't have to be a straight line. It would be more accurate to think of this situation as a "path integral."

5.11 Q: The path of a particle is described by the position vector $\vec{r} = <3t, t^2, -t>$. The particle experiences a force which varies as a function of the particle's position, described by the function $\vec{F} = <-2y, x^2, 3>$. Determine the work done on the particle by the force from time t=0 to t=3 seconds.

5.11 A: It is first important to note that your position and force functions are given in terms of different variables. The position function is given as a function of time, while the force function is given in terms of position of the particle. Note how you can determine the particle's x-, y-, and z-coordinates as functions of time from the position function.
$$x(t) = 3t \qquad y(t) = t^2 \qquad z(t) = -t$$

Now, you can re-write the force function as a function of time:
$$\vec{F} = <-2y, x^2, 3> \xrightarrow[x=3t]{y=t^2} \vec{F} = <-2t^2, 9t^2, 3>$$

Note that you'll have to find the derivative of $\vec{r}$ to complete your calculation.
$$\vec{r}(t) = <3t, t^2, -t> \to \frac{d\vec{r}}{dt} = <3, 2t, -1> \to d\vec{r} = <3, 2t, -1> dt$$

Next, you can integrate along the path to find the total work done by the force on the particle.
$$W = \int \vec{F} \cdot d\vec{r} = \int_{t=0}^{t=3} <-2t^2, 9t^2, 3> \cdot <3, 2t, -1> dt = \int_{t=0}^{t=3} (18t^3 - 6t^2 - 3) dt \to$$
$$W = \left(18\frac{t^4}{4} - 6\frac{t^3}{3} - 3 \right)\Big|_0^3 = \left(\frac{9}{2}t^4 - 2t^3 - 3t \right)\Big|_0^3 = (364.5 - 54 - 9) - 0 = 301.5\,J$$

Work-Energy Theorem

From the general work formula, you can derive an important formula known as the Work-Energy Theorem. Focusing on just a single dimension for simplicity, the work done by a force F in displacing an object from some initial position x_i to some final position x_f can be determined from

$$W = \int_{x_i}^{x_f} F(x)\,dx$$

You can then apply Newton's 2nd Law to replace F with the product of the object's mass and its acceleration, which is the product of the object's mass and the derivative of its velocity with respect to time.

$$W = \int_{x_i}^{x_f} F(x)\,dx \xrightarrow[v=\frac{dx}{dt} \to dx=vdt]{F=ma=m\frac{dv}{dt}} W = \int_{x_i}^{x_f} \left(m\frac{dv}{dt}\right)(vdt) = \int_{v_i}^{v_f} mvdv$$

Since the variable of integration changed from x to v, the limits of integration changed from x_i and x_f to v_i and v_f. Recognizing that the mass of the object is constant, you can continue with the integration.

$$W = \int_{v_i}^{v_f} mvdv = m\int_{v_i}^{v_f} vdv = \left(m\frac{v^2}{2}\right)\Big|_{v_i}^{v_f} = \tfrac{1}{2}mv_f^2 - \tfrac{1}{2}mv_i^2 \xrightarrow{K=\frac{1}{2}mv^2} W = K_f - K_i \to W = \Delta K$$

What does this tell you? The work done by the net force on an object is equal to the change in the object's kinetic energy! Work, therefore is the transfer of energy.

5.12 Q: A pickup truck with mass 1000 kg is traveling at 30 m/s. The driver sees a dog on the road 31 meters ahead. What force must the brakes exert in order to stop the truck in a distance of 30 meters, assuming constant acceleration?

5.12 A: Though there are multiple paths to solving this problem, we'll utilize the work-energy theorem to demonstrate its application.

$$W_{net} = \Delta K = K_f - K_i = \tfrac{1}{2}mv_f^2 - \tfrac{1}{2}mv_i^2 = 0 - \tfrac{1}{2}(1000)(30^2) = -450,000\,J$$

Knowing the net work, you can solve for the average force of the brakes.

$$W_{net} = F\Delta x \to F = \frac{W_{net}}{\Delta x} = \frac{450,000\,J}{30m} = -15,000\,N$$

5.13 Q: Given the sets of velocity and net force vectors for a given object, shown on the right, state whether you expect the kinetic energy of the object to increase, decrease, or remain the same.

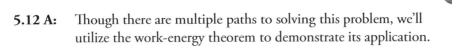

5.13 A: (A) decrease
(B) remain the same (no work done as long as v and F are perpendicular)
(C) increase
(D) decrease

5.14 Q: A chef pushes a 10-kilogram pastry cart from rest a distance of 5 meters with a constant horizontal force of 10 N. Assuming a frictionless surface, determine the cart's change in kinetic energy and its final velocity.

5.14 A: First find the work done by the chef, which will be equal to the cart's change in kinetic energy.
$$W = \vec{F} \cdot \vec{r} = (10N)(5m) = 50\,J$$

Next, solve for the cart's final velocity.
$$K = \tfrac{1}{2}mv^2 \rightarrow v = \sqrt{\frac{2K}{m}} = \sqrt{\frac{2(50J)}{10kg}} = 3.2\,^m\!/_s$$

5.15 Q: Given the force vs. displacement graph below for a net force applied horizontally to an object of mass m initially at rest on a frictionless surface, determine the object's final speed in terms of F_{max}, r_1, r_2, r_3, and m. You may assume the force does not change its direction.

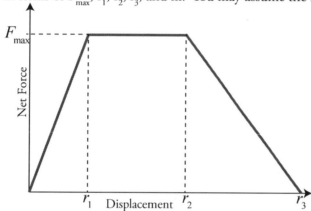

5.15 A: The work done is the area under the graph, which, according to the Work-Energy Theorem, is also equal to the change in the object's kinetic energy, therefore you can set the area of the graph equal to the object's final kinetic energy knowing the initial kinetic energy is zero.
$$W = Area = \tfrac{1}{2}bh + lw + \tfrac{1}{2}bh = \Delta K = \tfrac{1}{2}mv^2 \rightarrow$$
$$\tfrac{1}{2}r_1 F_{max} + (r_2 - r_1)F_{max} + \tfrac{1}{2}(r_3 - r_2)F_{max} = \tfrac{1}{2}mv^2 \rightarrow v^2 = \frac{F_{max}}{m}(r_3 + r_2 - r_1) \rightarrow$$
$$v = \sqrt{\frac{F_{max}}{m}(r_3 + r_2 - r_1)}$$

Energy and Conservative Forces

Although it has been touched on earlier, it's worth re-iterating as we dig deeper into the relationships between forces, energy, and work. If you recall, energy is the ability or capacity to do work, and work is the process of moving an object. Therefore, energy is the ability or capacity to move an object, and of course, it can be transformed from one type to another. You can transfer energy from one object to another by doing work. This allows you to re-state the Work-Energy Theorem as follows: "The work done on a system by an external force changes the energy of the system."

There are many different types of energy, which can be organized in various ways. Commonly, however, you'll find energy characterized as either kinetic or potential.

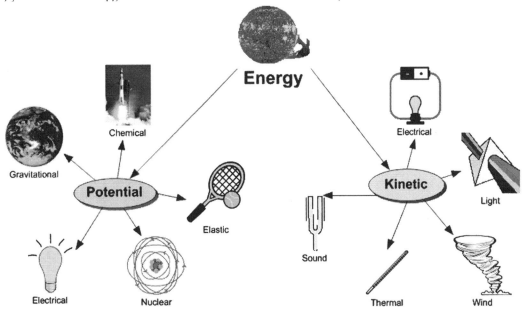

One type of energy that has been discussed is **kinetic energy**, or energy of motion. Kinetic energy is the ability or capacity of a moving object to move another object, and it can be calculated quite simply:

$$K = \tfrac{1}{2}mv^2$$

5.16 Q: A frog speeds along on his frog-o-cycle at a constant speed of 30 m/s. If the mass of the frog and the motorcycle is 5 kg, find the kinetic energy of the frog/cycle system.

5.16 A: $K = \tfrac{1}{2}mv^2 = \tfrac{1}{2}(5kg)\left(30\,^m\!/_s\right)^2 = 2250\,J$

Potential energy (U) is energy an object possesses due to its position or state of being. In many algebra-based courses, potential energy is symbolized with PE. In most upper-level courses, however, you'll find potential energy symbolized with U. A single object can have only kinetic energy, as potential energy requires an interaction between objects through forces. Some common types of potential energy discussed in this course include:

- **Gravitational Potential Energy** (U_g) is the energy an object possesses due to its position in a gravitational field.
- **Elastic Potential Energy** (U_s) is energy an object possesses due to the stretching or compressing of an elastic object such as a spring or rubber band.
- **Electrical Potential Energy** (U_E) is energy an electrically charged object possesses due to its position in an electric field.
- **Nuclear Potential Energy** (U_N) is energy the particles in the nucleus of an atom have due to strong and weak nuclear forces.

Take the example of a particle near the surface of the Earth in a region where the acceleration due to gravity is near constant. If it is initially at some position r_1, and over time follows a meandering path to its final point r_2, its y-position increases along the path by some amount h, as shown in the diagram at right. You can calculate the work done on the object from the general definition of work.

$$W = \int_{r_1}^{r_2} \vec{F} \cdot d\vec{r}$$

The force needed to lift the object up at constant speed must be equal to the weight of the object, mg, and the dot product of the force (upward) and the differential of the path leaves you with just the differential of the path in the y-direction. You can therefore re-write the definition and calculate the integral.

$$W = \int_{r_1}^{r_2} \vec{F} \cdot d\vec{r} = \int_{y=h_i}^{y=h_f} mg\, dy = mg \int_{y=h_i}^{y=h_f} dy = mgh \Big|_{h_i}^{h_f} = mg\Delta h$$

You have therefore done $mg\Delta h$ amount of work to move the object from r_1 to r_2, and as gravity is a conservative force (which will be discussed shortly), the gravitational potential energy of the object has increased by an amount $mg\Delta h$. In a constant gravitational field, the gravitational potential energy of an object with respect to some reference point can be written as:

$$U_g = mgh$$

5.17 Q: The diagram below represents a 155-newton box on a ramp. Applied force F causes the box to slide from point A to point B.

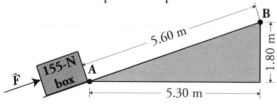

What is the total amount of gravitational potential energy gained by the box?
(A) 28.4 J
(B) 279 J
(C) 868 J
(D) 2740 J

5.17 A: (B) $\Delta U_g = mg\Delta y = (155N)(1.80m) = 279\ J$

5.18 Q: A hippopotamus is thrown vertically upward. Which pair of graphs best represents the hippo's kinetic energy and gravitational potential energy as functions of its displacement while it rises?

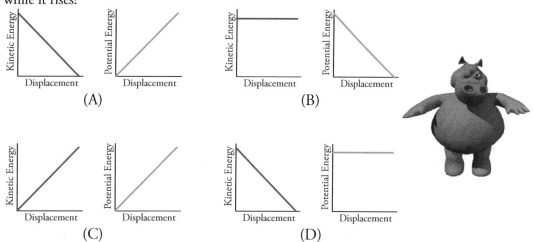

(A)

(B)

(C)

(D)

5.18 A: (A) shows the hippo's kinetic energy decreasing as it slows down on its way upward, while its potential energy increases as its height increases.

5.19 Q: A pendulum of mass M swings on a light string of length L as shown in the diagram below right. If the mass hanging directly down is set as the zero point of gravitational potential energy, find the gravitational potential energy of the pendulum as a function of θ and L.

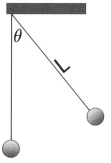

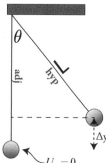

5.19 A: First use some basic geometry and trigonometry to analyze the situation and set up the problem. Re-draw the diagram as shown above right, then solve for the change in vertical displacement, Δy.

$$\cos\theta = \frac{adj}{hyp} \rightarrow adj = hyp \times \cos\theta = L\cos\theta$$

$$\Delta y = L - adj = L - L\cos\theta = L(1-\cos\theta)$$

Now, utilize your value for Δy to find the gravitational potential energy above your reference point.

$$\Delta U_g = mg\Delta y \xrightarrow{\Delta y = L(1-\cos\theta)} \Delta U_g = mgL(1-\cos\theta)$$

A force in which the work done on an object is independent of the path taken is known as a **conservative force**. For example, if you lift a book straight upward, or if you move it in a bunch of vertical circles, as long as you start at the same point and end at the same point, you have done the same amount of work on the book. This indicates that gravity is a conservative force. You could also state that the work done in moving along a closed path is zero. Regardless of the path taken, if you start and end at the same position, the net work done on the book is zero. Finally, you can define a conservative force as a force in which the work done is directly related to a negative change in potential energy.

Some examples of conservative forces include forces such as the gravitational force, elastic force, and electrical forces. Non-conservative forces are forces you tend to think of as "lossy" forces, and include forces such as friction, air resistance, and fluid resistance.

The work done by a conservative force is equal to the opposite of the change in the object's potential energy. Written mathematically:

$$W_{conservative\ force} = -\Delta U \rightarrow \Delta U = -W_{cons.f} \rightarrow \Delta U = -\int_{r_1}^{r_2} \vec{F} \cdot d\vec{r}$$

Exploring elastic potential energy is a great demonstration of this conservative force / potential energy relationship. Consider a spring that obeys Hooke's Law, where $\vec{F}_s = -k\vec{x}$. The work done in stretching the spring can be determined as:

$$W = -\Delta U_s = -\int_0^x \vec{F} \cdot d\vec{r} = -\int_0^x -kxdx \rightarrow U_s = k\int_0^x xdx = \tfrac{1}{2}kx^2$$

You have therefore derived the formula for the potential energy stored in a linear spring.

If you can find potential energy from force, it only makes sense that you should be able to find force when you know the potential energy function. Imagine a small displacement $d\vec{l}$ of an object moving along a path under the influence of a conservative force $\vec{F}$ as shown below.

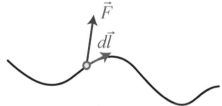

If you wanted to find the potential energy due to the force along the path, you can break the path up into tiny pieces dl, find the work done by the force in each tiny section, then add up the work done in each of the tiny sections. Written mathematically

$$dU = -dW_F = -\vec{F} \cdot d\vec{l} = -F\cos\theta dl \xrightarrow{F_l = F\cos\theta} dU = -F_l dl \rightarrow F_l = -\frac{dU}{dl}$$

This states that the differential of the change in potential energy is the opposite of the differential of the work done by the conservative force, which you can calculate using the dot product. If you define a quantity F_l as the component of the force in the direction of the displacement, you can re-arrange your equation to find that the component of the force in the direction of the displacement, F_p is the opposite of the derivative of the potential energy function with respect to l. Now you have a method of finding force from the potential energy function!

Since you already determined the elastic potential energy stored in a spring from Hooke's Law, let's use this relationship to start with the spring elastic potential energy and determine the force of a spring (Hooke's Law).

$$-\frac{dU}{dl} = F_l \xrightarrow{U_s = \frac{1}{2}kx^2} -\frac{d}{dx}\left(\tfrac{1}{2}kx^2\right) = F_s \rightarrow F_s = -kx$$

5.20 Q: Given the gravitational potential energy near the surface of the Earth is given by $U_g = mgh$, find the gravitational force.

5.20 A: $F = -\dfrac{dU}{dl} \rightarrow F_g = -\dfrac{dU_g}{dl} = -\dfrac{d}{dh}mgh = -mg$. Note that the gravitational force is negative, indicating the force (down) is pointing opposite the direction of increasing gravitational potential energy (up).

5.21 Q: A 5-kg sphere's position is given by x(t) = 3t³-2t. Determine the kinetic energy of the sphere at time t=3 seconds.

5.21 A: $K = \tfrac{1}{2}mv^2 \xrightarrow[v(t=3s)=9(3)^2-2=79\,m/s]{x(t)=3t^3-2t,\, v(t)=x'(t)=9t^2-2} K = \tfrac{1}{2}(5kg)(79\,m/s)^2 = 15,600\,J$

5.22 Q: Another 5-kg sphere's potential energy U is described by U(x)=4x²+3x-2. Determine the force on the particle at x=2 meters.

5.22 A: $F = -\dfrac{dU}{dl} = -\dfrac{d}{dx}\left(4x^2 + 3x - 2\right) = -(8x+3) \xrightarrow{x=2m} F(2m) = -(8(2)+3) = -19N$

5.23 Q: A 2-kg disc moves in uniform circular motion on a frictionless horizontal table, attached to the point of rotation by a 10-cm spring with a spring constant of 50 N/m as shown below.

When stationary, the spring has a length of 8 cm. How much work is performed on the disc by the spring as the disc moves through one full revolution.

5.23 A: No work is done by the spring on the disc because the force is perpendicular to the displacement.

Often times you may hear the term **internal energy** when discussing systems in physics. The internal energy of a system includes the kinetic energy of the objects that make up the system and the potential energy of the configuration of the objects that make up the system. Changes in a system's internal structure can result in changes in internal energy. For example, as a book slides across a desk, the molecules of the desk move more quickly, which is a change in the average kinetic energy of the molecules of the desk (what we perceive as an increase in the desk's temperature). This is an example of an increase in the desk's internal energy.

Conservation of Energy

"Energy cannot be created or destroyed... it can only be changed."

Chances are you've heard that phrase before. It's one of the most important concepts in all of physics. It doesn't mean that an object can't lose energy or gain energy. What it means is that energy can be changed into different forms, and transferred from system to system, but it never magically disappears or reappears. In the world of physics, you can never truly destroy energy. The understanding that the total amount of energy in the universe remains fixed is known as the **law of conservation of energy**.

Objects and systems can possess multiple types of energy. The energy of a system includes its kinetic energy, potential energy, and its internal energy. **Mechanical energy** is the sum of an object's kinetic energy as well as its gravitational potential and elastic potential energies. Non-mechanical energy forms include chemical potential, nuclear, and thermal.

Consider a single conservative force doing work on a closed system. The work done by the force must be the change in the kinetic energy, consistent with the work-energy theorem. You also know the work done by a conservative force is equal to the opposite of the change in the object's potential energy:

$$W_{conservative\ force} = \Delta K \xrightarrow{W_{conservative\ force} = -\Delta U} \Delta K = -\Delta U \rightarrow \Delta K + \Delta U = 0$$

This indicates that the total change of energy in the closed system must be zero; therefore, the initial energy of the system must equal the final energy of the system, and if dealing with only mechanical energy, the sum of the initial kinetic and potential energies must equal the sum of the final kinetic and potential energies.

$$\Delta K + \Delta U = 0 \rightarrow E_i = E_f \rightarrow K_i + U_i = K_f + U_f$$

Non-conservative forces change the total mechanical energy of a system, but not the total energy of the system. The work done by a non-conservative force is typically converted to internal (thermal) energy, though this is not necessarily the case.

5.24 Q: An object of mass m falls from a height h. Find its speed immediately prior to impact using conservation of mechanical energy. Neglect air resistance and assume a constant gravitational field.

5.24 A: $E_i = E_f \rightarrow K_i + U_i = K_f + U_f \xrightarrow[K_f = \frac{1}{2}mv_f^2, U_f = 0]{U_i = mgh, K_i = 0} mgh = \frac{1}{2}mv_f^2 \rightarrow v_f = \sqrt{2gh}$

5.25 Q: A jet fighter with a mass of 20,000 kg coasts through the sky at an altitude of 10 km and a velocity of 250 m/s. The jet then dives to an altitude of 2 km. Find the new velocity of the jet. Assume all forces acting on the system are conservative.

5.25 A: If all the forces acting on the system are conservative, mechanical energy must be conserved.

$$E_i = E_f \rightarrow K_i + U_i = K_f + U_f \rightarrow mgh_i + \tfrac{1}{2}mv_i^2 = mgh_f + \tfrac{1}{2}mv_f^2 \rightarrow$$

$$v_f = \sqrt{2g\Delta h + v_i^2} \xrightarrow[\substack{\Delta h = 8000m \\ v_i = 250^m/_s}]{} v_f = \sqrt{2\left(10^m/_{s^2}\right)\left(8000m\right) + \left(250^m/_s\right)^2} = 472^m/_s$$

An ideal pendulum is a common device for demonstrating conservation of mechanical energy. Assume a mass M swings back and forth on a light (massless) string of length L, as shown below.

As the pendulum transitions from its highest points to its lowest point, energy transforms from gravitational potential energy to kinetic energy.

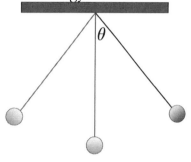

The difference in height from the highest to lowest point can be determined using basic trigonometry, as you've done previously, allowing you to calculate the maximum speed of the mass in terms of the string length and the maximum angular displacement of the mass from the vertical by recognizing that the maximum kinetic energy, at the lowest point, is equal to the change in gravitational potential energy at the highest point.

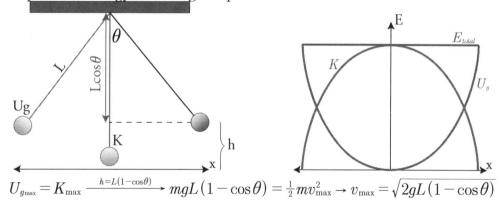

$$U_{g\,max} = K_{max} \xrightarrow{\;h = L(1-\cos\theta)\;} mgL\left(1 - \cos\theta\right) = \tfrac{1}{2}mv_{max}^2 \rightarrow v_{max} = \sqrt{2gL\left(1 - \cos\theta\right)}$$

5.26 Q: A toy cart possessing 16 joules of kinetic energy travels on a frictionless, horizontal surface toward a horizontal spring. If the cart comes to rest after compressing the spring a distance of 1 meter, what is the spring constant of the spring?

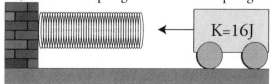

5.26 A: Utilize conservation of energy by recognizing that the cart's initial kinetic energy must be equal to the potential energy in the compressed spring once the cart comes to a stop.

$$K_i = U_{s_f} \rightarrow 16J = \tfrac{1}{2}kx^2 \rightarrow k = \frac{32J}{(1m)^2} = 32\,{}^N\!/\!_m$$

5.27 Q: A pop-up toy has a mass of 20 grams and a spring constant of 150 N/m. A force is applied to the toy to compress the spring 5 cm.
A) Calculate the potential energy stored in the compressed spring.
B) Determine the maximum vertical height to which the toy can be propelled.

5.27 A: A) $U_s = \tfrac{1}{2}kx^2 = \tfrac{1}{2}(150\,{}^N\!/\!_m)(0.05m)^2 = 0.1875\ J$

B) $U_i = U_f \rightarrow 0.1875\ J = mgh \rightarrow h = \dfrac{0.1875J}{(0.02kg)(10\,{}^m\!/\!_{s^2})} = 0.9375m$

5.28 Q: A car, initially traveling at 30 m/s, slows uniformly as it skids to a stop after the brakes are applied. Sketch a graph showing the relationship between the kinetic energy of the car as it is being brought to a stop and the work done by friction in stopping the car.

5.28 A:

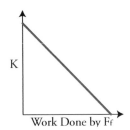

5.29 Q: A 2-kilogram block sliding down a ramp from a height of 3 meters above the ground reaches the ground with a kinetic energy of 50 joules. Find the total work done by friction on the block as it slides down the ramp.

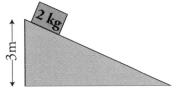

5.29 A: The gravitational potential energy of the block at the top of the ramp must equal the sum of the block's kinetic energy at the bottom of the ramp and the work done by friction.
$$U_{g_{top}} = K_{bottom} + W_{friction} \rightarrow W_{friction} = U_{g_{top}} - K_{bottom} = mgh_{top} - 50J \rightarrow$$

$$W_{friction} = (2kg)(10\,{}^m\!/\!_{s^2})(3m) - 50J = 10\ J$$

Chapter 5: Work, Energy, and Power

5.30 Q: Andy the adventurous adventurer, while running from headhunters deep in the rainforest, trips, falls, and slides down a mudslide of height 20 meters, as depicted below. Once he reaches the bottom of the mudslide, he has the misfortune of flying horizontally off a 15-meter cliff. How far from the base of the cliff does Andy land? Neglect friction.

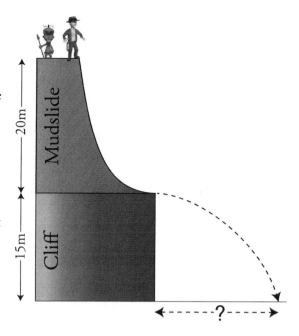

5.30 A: This problem can be attacked in multiple steps. First, analyze just the mudslide part of the problem utilizing conservation of energy to find the speed with which Andy flies horizontally off the cliff.

$$U_{g_{top}} = K_{bottom} \rightarrow mgh = \tfrac{1}{2}mv^2 \rightarrow$$
$$v = \sqrt{2gh} = \sqrt{2(10)(20)} = 20\,{}^m\!/_s$$

Next, you can analyze Andy's projectile motion using your 2-D kinematics skills to find Andy's horizontal displacement. Start by finding the amount of time Andy is in the air by analyzing the vertical motion.

$$\Delta y = v_{0_y} + \tfrac{1}{2}a_y t^2 \xrightarrow{\ v_{0_y}=0\ } t = \sqrt{\frac{2\Delta y}{a_y}} = \sqrt{\frac{2(15m)}{10\,{}^m\!/_{s^2}}} = 1.73s$$

Next, find Andy's horizontal displacement, having already solved for his initial velocity by analyzing the mudslide section of the problem.

$$\Delta x = v_{0_x}t = (20\,{}^m\!/_s)(1.73s) = 34.6m$$

5.31 Q: Two children on the playground, Bobby and Sandy, travel down frictionless slides of identical height h but different shapes as shown at right. Assuming they start from rest and begin their journey down the slides at the same time, which of the following statements is true?
(A) Bobby reaches the bottom first with the same average velocity as Sandy.
(B) Bobby reaches the bottom first with a larger average acceleration than Sandy.
(C) Bobby reaches the bottom first with the same average acceleration as Sandy.
(D) They reach the bottom at the same time with the same average acceleration.

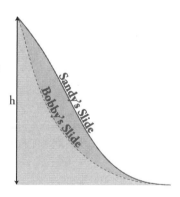

5.31 A: (B) Both children begin with gravitational potential energy mgh at the top of the slide, which is completely transferred to kinetic energy at the end of the slide. Bobby's potential energy is transferred more quickly, however; therefore, he attains a higher average velocity and beats Sandy to the end of the slide. Average acceleration is the change in velocity divided by the time interval. Each child has the same change in velocity, but Bobby observes this change over a shorter period of time, resulting in a larger average acceleration.

5.32 Q: A net force ($F\cos\theta$) acts on an object in the x-direction while moving over a distance of 4 meters along the axis, depicted in the graph at right.

A) Find the work done by the force from 0.0 to 1.0 m.
B) Find the work done by the force from 1.0 to 2.0 m.
C) Find the work done by the force from 2.0 to 4.0 m.
D) At what position(s) is the object moving with the largest speed? Explain your answer.

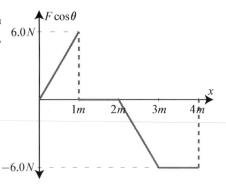

5.32 A: A) 3 J
B) 0 J
C) -9 J
D) The object is moving with the largest speed at the positions of 1 to 2 m. From 0 to 1 m the force is increasing the speed of the object where it reaches a maximum value. From 1 to 2 m there is no force acting on the object therefore it is moving at a constant speed. From 2 to 4 m the negative force is slowing the object down since the object is still moving in the positive x direction.

5.33 Q: If you throw a rock straight up outside, it eventually returns to your hand with the same speed that it had when it left, neglecting air resistance. What would happen if you were to throw the rock straight up in the same way, but while inside the classroom? Compared to the speed with which it left your hand, after rebounding off of the ceiling it would return to your hand with...
(A) a higher speed.
(B) a lower speed.
(C) the same speed, just like when you are outside.
(D) more information needed
Explain your reasoning.

5.33 A: (B) a lower speed. Analyzing this problem through the lens of conservation of energy, some energy is lost in the collision to sound, internal energy, work done on the ceiling, etc., so when the rock returns to your hand, it has to have less energy than it began with.

5.34 Q: Bowling Ball A is dropped from a point halfway up a cliff. A second identical bowling ball, B, is dropped simultaneously from the top of the cliff. Comparing the bowling balls at the instant they reach the ground, which of the following are correct? Neglect air resistance.
(A) Ball A has half the kinetic energy and takes half the time to hit the ground as Ball B.
(B) Ball A has half the kinetic energy and takes one-fourth the time to hit the ground as Ball B.
(C) Ball A has half the final velocity and takes half the time to hit the ground as Ball B.
(D) Ball A has one-fourth the final velocity and takes one-fourth the time to hit the ground as Ball B.
(E) None of these are correct.

5.34 A: (E) The final velocities of the balls are given by $v = \sqrt{2gh}$. The final velocity of B is related to the square root of the height, therefore the final velocity of B is $\sqrt{2}$ times the final velocity of A, eliminating choices C and D. Further, the kinetic energy of Ball A is half the kinetic energy of Ball B at the instant the balls reach the ground. The time it takes for the balls to reach the ground is also related to the square root of the height, therefore the time for B to hit the ground is $\sqrt{2}$ times the time for A to hit the ground, eliminating choices A and B. E must be the correct answer.

5.35 Q: Identical metal blocks initially at rest are released in various environments as shown in scenarios A through D below. In all cases, the blocks are released from a height of 2m. Neglect air resistance.

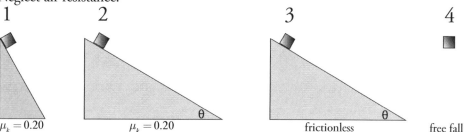

Rank the scenarios from least kinetic energy to greatest kinetic energy at the instant before the block reaches the ground.
(A) 2 < 3 < 1 < 4
(B) 2 < 1 < 3 = 4
(C) 2 < 1 < 3 < 4
(D) 1 = 2 < 3 = 4
(E) 2 < 3 = 1 < 4

5.35 A: (B) 2 < 1 < 3 = 4. Scenarios 3 and 4 both start at the same height, with the same gravitational potential energy, and without friction, they must have the same mechanical energy at height 0, therefore they have the same kinetic energy. Scenarios 1 and 2 have both lost some energy to friction, so they will have a smaller amount of kinetic energy at the bottom. At a greater (steeper) angle, scenario 1 has a smaller frictional force over a shorter distance, therefore less work is done by friction, leaving more kinetic energy at the bottom.

5.36 Q: A roller coaster car begins at rest at height h above the ground and completes a loop along its path. In order for the car to remain on the track throughout the loop, what is the minimum value for h in terms of the radius of the loop, R? Assume frictionless.

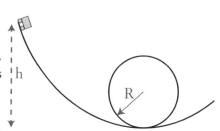

5.36 A: Use the law of conservation of energy to find the speed of the cart at the top of the loop.
$$U_i = K_f + U_f \rightarrow mgh = \tfrac{1}{2}mv_f^2 + mg(2R) \rightarrow 2gh = v_f^2 + 4gR \rightarrow v_f^2 = 2g(h - 2R)$$

Next, recognize for the cart to stay on the loop and not fall off, the centripetal acceleration must be equal to or greater than the acceleration due to gravity.
$$\frac{v^2}{R} \geq g \rightarrow v^2 \geq gR \rightarrow 2g(h - 2R) \geq gR \rightarrow h - 2R \geq \frac{R}{2} \rightarrow h \geq \frac{5R}{2}$$

Power

Power is a term used quite regularly in all aspects of life. People talk about how powerful the new boat motor is, the power of positive

thinking, and even the power company's latest bill. All of these uses of the term power relate to how much work can be done in some amount of time, or the rate at which energy is transferred.

In physics, work can be defined in two ways. Work is the process of moving an object by applying a force. The rate at which the force does work is known as **power** (P). Work is also the transfer of energy, so power is also the rate at which energy is transferred into, out of, or within a system. The units of power are the units of work divided by time, or joules per second, known as a watt (W).

$$P_{avg} = \frac{\Delta W}{\Delta t}$$

You can determine the instantaneous power delivered by shrinking the time interval down to an infinitessimally small amount of time, effectively taking the derivative of the work done with respect to time.

$$P = \frac{dW}{dt}$$

Recognizing that the differential of work can be found from $dW = \vec{F} \cdot d\vec{r}$, you can manipulate this equation further to find power as a function of the force and velocity vectors.

$$P = \frac{dW}{dt} \xrightarrow{dW = \vec{F} \cdot d\vec{r}} P = \frac{\vec{F} \cdot d\vec{r}}{dt} \xrightarrow{\frac{d\vec{r}}{dt} = \vec{v}} P = \vec{F} \cdot \vec{v}$$

So, not only is power equal to work done divided by the time required, it's also equal to the dot product of the force and velocity vectors!

5.37 Q: Santa drags a Christmas tree across a horizontal surface at a constant speed of 1 m/s. If the tree has a mass of 30 kg, find the power Santa supplies given the coefficient of kinetic friction between the tree and the ground is 0.3.

5.37 A: First draw a free body diagram for the tree. You can then solve for the force of Santa by applying Newton's 2nd Law in both the x- and y- directions.

FBD

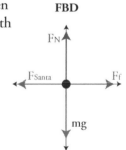

$$F_{net_x} = F_{f_k} - F_{Santa} = ma_x \xrightarrow{a_x = 0} F_{Santa} = F_{f_k} = \mu_k F_N \xrightarrow[F_N = mg]{F_{net_y} = F_N - mg = 0}$$

$$F_{Santa} = \mu_k mg = (0.3)(30kg)(10\tfrac{m}{s^2}) = 90N$$

Next you can calculate the power supplied, recognizing the force and velocity vectors are in the same direction, therefore the angle between them is zero.

$$P = \vec{F} \cdot \vec{v} = Fv \cos\theta \xrightarrow{\theta = 0} P = Fv = (90N)(1\tfrac{m}{s}) = 90W$$

5.38 Q: A 2000-kg truck starts from rest and "moo"ves in a straight line for 12 seconds with a velocity function given by v(t)=4t. What is the average power required to accomplish this?

5.38 A: Recalling that $P_{avg} = F_{avg} \cdot v_{avg}$, you can find the average force using Newton's 2nd Law and a bit of calculus.

$$F_{avg} = ma_{avg} \xrightarrow[a(t)=\frac{dv(t)}{dt}=4]{v(t)=4t} (2000kg)(4\,^m\!/_{s^2}) = 8000N$$

Next, you can find the average velocity by taking the displacement divided by the time interval.

$$v_{avg} = \frac{\Delta x}{t} \xrightarrow{\Delta x=\int_{t=0}^{12s} v(t)dt = \int_{t=0}^{12s} 4t\,dt = 2t^2\big|_0^{12s} = 288m} v_{avg} = \frac{288m}{12s} = 24\,^m\!/_s$$

Finally, solve for the average power.

$$P_{avg} = F_{avg} \cdot v_{avg} = (8000N)(24\,^m\!/_s) = 192,000\,W = 192\,kW$$

5.39 Q: A box of mass m is pushed up a ramp at constant velocity v to a maximum height h in time t by force F, as shown at right. The ramp makes an angle of θ with the horizontal as shown in the diagram. What is the power supplied by the force? Choose all that apply.

A) $\dfrac{mgh}{t}$

B) $\dfrac{mgh}{t\sin\theta}$

C) $\dfrac{Fh}{t\sin\theta}$

D) Fv

5.39 A: C & D. Answers A and B are variations of a solution based off the idea of using the work done divided by the time, which only holds if the problem states that there is no friction. Since the problem does not state the ramp is frictionless, you can solve for power using $P = Fv$. This can be extended with a bit of geometry and trigonometry to give you answer C as well. $P = Fv \xrightarrow[v=\frac{d}{t}=\frac{h}{t\sin\theta}]{d=\frac{h}{\sin\theta}} P = \dfrac{Fh}{t\sin\theta}$

5.40 Q: Find the power delivered by the net force to a 10-kg mass at time t=4s given the position of the mass is given by x(t)=4t³-2t.

5.40 A: Use $P = \vec{F} \cdot \vec{v}$. This will require you to first find the velocity and acceleration functions, as well as the net force on the mass.

$$v(t) = \frac{dx(t)}{dt} = 12t^2 - 2, \quad a(t) = \frac{dv(t)}{dt} = 24t, \quad F_{net} = ma = (10)(24t) = 240t$$

$$P = \vec{F} \cdot \vec{v} = Fv\cos\theta \xrightarrow[\cos\theta=1]{\theta=0} P = (240t)(12t^2 - 2) = 2880t^3 - 480t \xrightarrow{t=4s}$$

$$P = 182,400\,W = 182\,kW$$

Chapter 6: Momentum

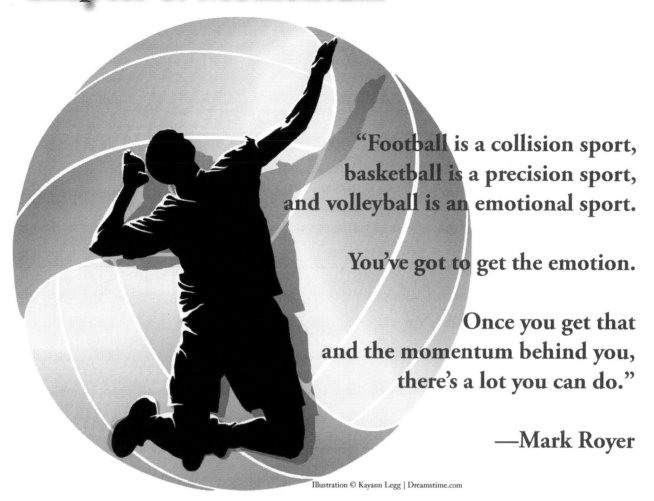

"Football is a collision sport, basketball is a precision sport, and volleyball is an emotional sport.

You've got to get the emotion.

Once you get that and the momentum behind you, there's a lot you can do."

—Mark Royer

Illustration © Kayann Legg | Dreamstime.com

Objectives

1. Calculate the total linear momentum of a system of objects.

2. Describe the relationship between forces acting on an object, impulse, and changes in linear momentum.

3. Apply the relationship between linear momentum and center of mass motion for a system of particles.

4. Calculate the change in momentum of an object given a force acting on the object as a function of time.

5. Apply conservation of linear momentum to collisions and explosions in multiple dimensions.

6. Identify the center-of-mass of a symmetrical object by inspection.

7. Use integration to determine the center of mass of various objects with both uniform and non-uniform density functions.

8. Understand and utilize the concepts of center of gravity and center of mass to find the gravitational potential energy of an object.

You've explored motion in some depth, specifically trying to relate what you know about motion back to kinetic energy. Recall the definition of kinetic energy as the ability or capacity of a moving object to move another object. The key characteristics of kinetic energy, mass and velocity, can be observed from the equation:

$$K = \tfrac{1}{2}mv^2$$

There's more to the story, however. Moving objects may cause other objects to move, but these interactions haven't been explored yet. To learn more about how one object causes another to move, you need to learn about collisions, and collisions are all about momentum.

Defining Momentum

Assume there's a car speeding toward you, out of control without its brakes, at a speed of 27 m/s (60 mph). Can you stop it by standing in front of it and holding out your hand? Why not?

Unless you're Superman, you probably don't want to try stopping a moving car by holding out your hand. It's too big, and it's moving way too fast. Attempting such a feat would result in a number of physics demonstrations upon your body, all of which would hurt.

You can't stop the car because it has too much momentum. **Momentum** is a vector quantity, given the symbol $\vec{p}$, which measures how hard it is to stop a moving object. Of course, larger objects have more momentum than smaller objects, and faster objects have more momentum than slower objects. You can therefore calculate momentum using the equation:

$$\vec{p} = m\vec{v}$$

Momentum is the product of an object's mass times its velocity, and its units must be the same as the units of mass [kg] times velocity [m/s]; therefore, the units of momentum must be [kg·m/s], which can also be written as a newton-second [N·s].

6.01 Q: Two trains, Big Red and Little Blue, have the same velocity. Big Red, however, has twice the mass of Little Blue. Compare their momenta.

6.01 A: Because Big Red has twice the mass of Little Blue, Big Red must have twice the momentum of Little Blue.

6.02 Q: The magnitude of the momentum of an object is 64 kilogram-meters per second. If the velocity of the object is doubled, the magnitude of the momentum of the object will be
(A) 32 kg·m/s
(B) 64 kg·m/s
(C) 128 kg·m/s
(D) 256 kg·m/s

6.02 A: (C) if velocity is doubled, momentum is doubled.

Because momentum is a vector, the direction of the momentum vector is the same as the direction of the object's velocity.

6.03 Q: An Aichi D3A bomber, with a mass of 3600 kg, departs from its aircraft carrier with a velocity of 85 m/s due east. What is the plane's momentum?

6.03 A: $\vec{p} = m\vec{v} = (3600 kg)(85 \, ^m/_s \, \hat{i}) = 3.06 \times 10^5 \, ^{kg \cdot m}/_s \, \hat{i}$

Now, assume the bomber drops its payload and has burned up most of its fuel as it continues its journey east to its destination air field.

6.04 Q: If the bomber's new mass is 3,000 kg, and due to its reduced weight the pilot increases the cruising speed to 120 m/s, what is the bomber's new momentum?

6.04 A: $\vec{p} = m\vec{v} = (3000 kg)(120 \, ^m/_s \, \hat{i}) = 3.60 \times 10^5 \, ^{kg \cdot m}/_s \, \hat{i}$

6.05 Q: Cart A has a mass of 2 kilograms and a speed of 3 meters per second. Cart B has a mass of 3 kilograms and a speed of 2 meters per second. Compared to the inertia and magnitude of momentum of cart A, cart B has
(A) the same inertia and a smaller magnitude of momentum
(B) the same inertia and the same magnitude of momentum
(C) greater inertia and a smaller magnitude of momentum
(D) greater inertia and the same magnitude of momentum

6.05 A: (D) greater inertia and the same magnitude of momentum.

Force and Momentum

When Isaac Newton formulated the second law of motion in his Principia, he didn't state it in terms of force equals the product of mass and acceleration. Instead, Newton stated that the rate at which an object's momentum changes is directly proportional and in the same direction as the applied force. Written mathematically:

$$\vec{F} = \frac{d\vec{p}}{dt} \xrightarrow{\vec{p}=m\vec{v}} \vec{F} = \frac{d}{dt}(m\vec{v})$$

This says that when a force acts on a particle, it changes the particle's momentum, and if a particle's momentum changes, a force must have acted on it.

When you assume a constant-mass system, where Newton's 2nd Law is valid, you derive the more commonly written version.

$$\vec{F} = \frac{d\vec{p}}{dt} \xrightarrow{\vec{p}=m\vec{v}} \vec{F} = \frac{d}{dt}(m\vec{v}) \xrightarrow{m \ constant} \vec{F} = m\frac{d\vec{v}}{dt} \rightarrow \vec{F} = m\vec{a}$$

6.06 Q: The magnitude of the momentum of an object as a function of time is given by p(t)=kt². What is magnitude of the force causing this motion?

6.06 A: $F = \dfrac{dp}{dt} = \dfrac{d}{dt}(kt^2) = k\dfrac{d}{dt}(t^2) = 2kt$

6.07 Q: Mass m_1 is held in place on a frictionless table, and is connected by a light string across a massless pulley to mass m_2 as shown below.

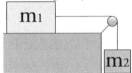

What is the magnitude of the total momentum of the boxes and string when the speed of m_1 is v?
A) $m_1 v + m_2 v$
B) $\frac{1}{2}(m_1 + m_2)v^2$
C) $v^2\sqrt{(m_1 + m_2)}$
D) $\sqrt{m_1 v} + \sqrt{m_2 v}$
E) $v\sqrt{(m_1^2 + m_2^2)}$

6.07 A: E) $v\sqrt{(m_1^2 + m_2^2)}$. Recall that momentum is a vector. Assuming to the right is the positive x-direction and down is the positive y-direction, when the speed of m_1 is v, its velocity is $v\hat{i}$. The velocity of m_2 at that time is $v\hat{j}$. The momentum of the system at that time must be $\vec{p} = m_1 v\hat{i} + m_2 v\hat{j}$.

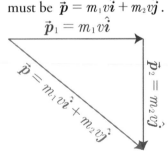

Find the magnitude of the total momentum using the Pythagorean Theorem:
$$|\vec{p}| = \sqrt{(m_1 v)^2 + (m_2 v)^2} = \sqrt{m_1^2 v^2 + m_2^2 v^2} = v\sqrt{(m_1^2 + m_2^2)}$$

6.08 Q: The force on an object as a function of time is given by the equation F(t)=6t+3. Determine the change in the object's momentum from t=0 to t=3 seconds.

Chapter 6: Momentum

6.08 A: $F = \dfrac{dp}{dt} \rightarrow dp = Fdt \rightarrow \displaystyle\int_{p_i}^{p_f} dp = \int_{t=0}^{3s} Fdt \xrightarrow{F(t)=6t+3} p\,\big|_{p_i}^{p_f} = \int_{t=0}^{3s} (6t+3)\,dt \rightarrow$

$p_f - p_i = (3t^2 + 3t)\,\big|_{t=0}^{3s} \xrightarrow{p_f - p_i = \Delta p} \Delta p = 36\ N \cdot s$

Impulse

As you know, the momentum of an object can change. A change in momentum is known as an **impulse**. In physics, the vector quantity impulse is represented by $\vec{J}$ and since it's a change in momentum, its units are the same as those for momentum, [kg·m/s], which can also be written as a newton-second [N·s].

$$\vec{J} = \Delta\vec{p}$$

6.09 Q: Calculate the magnitude of the impulse applied to a 0.75-kilogram cart to change its velocity from 0.50 meter per second east to 2.00 meters per second east.

6.09 A: $J = \Delta p = m\Delta v = (0.75kg)(1.5\,{}^m\!/_s) = 1.1\ N \cdot s$

6.10 Q: A 6.0-kilogram block, sliding to the east across a horizontal, frictionless surface with a momentum of 30 kilogram•meters per second, strikes an obstacle. The obstacle exerts an impulse of 10 newton•seconds to the west on the block. The speed of the block after the collision is
(A) 1.7 m/s
(B) 3.3 m/s
(C) 5.0 m/s
(D) 20 m/s

6.10 A: (B) $J = \Delta p = m\Delta v = m(v - v_0) \rightarrow v = \dfrac{J + mv_0}{m} = \dfrac{(-10N \cdot s) + 30\,{}^{kg \cdot m}\!/_s}{6kg} = 3.3\,{}^m\!/_s$

6.11 Q: A 1000-kilogram car traveling due east at 15 meters per second is hit from behind and receives a forward impulse of 6000 newton-seconds. Determine the magnitude of the car's change in momentum due to this impulse.

6.11 A: Change in momentum is the definition of impulse; therefore, the answer must be 6000 newton-seconds.

Since momentum is equal to mass times velocity, you can write that p=mv. Since you also know impulse is a change in momentum, impulse can be written as J=Δp. Combining these equations, you find:

$$\vec{J} = \Delta\vec{p} = \Delta m\vec{v}$$

Since the mass of a single object is constant, a change in the product of mass and velocity is equivalent to the product of mass and change in velocity. Specifically:

$$\vec{J} = \Delta\vec{p} = m\Delta\vec{v}$$

A change in velocity is called acceleration. But what causes an acceleration? A force! And does it matter if the force is applied for a very short time or a very long time? Common sense says it does and also tells us that the longer the force is applied, the longer the object will accelerate, and therefore the greater the object's change in momentum!

You can prove this using an old mathematician's trick. If you multiply the right side of the equation by 1, you of course get the same thing. And if you multiply the right side of the equation by $\Delta t/\Delta t$, which is 1, you still get the same thing. Take a look:

$$\vec{J} = \Delta\vec{p} = \frac{m\Delta\vec{v}\Delta t}{\Delta t}$$

If you look carefully at this equation, you can find a $\Delta v/\Delta t$, which is, by definition, acceleration. By replacing $\Delta v/\Delta t$ with acceleration a in the equation, you arrive at:

$$\vec{J} = \Delta\vec{p} = m\vec{a}\Delta t$$

One last step... perhaps you can see it already. On the right-hand side of this equation, you have $ma\Delta t$. Utilizing Newton's 2nd Law, you can replace the product of mass and acceleration with force F, giving the final form of the equation, oftentimes referred to as the **Impulse-Momentum Theorem**:

$$\vec{J} = \Delta\vec{p} = \vec{F}\Delta t$$

You can also derive this directly from Newton's statement of the second law of motion.

$$\vec{F} = \frac{d\vec{p}}{dt} \rightarrow \int_0^t \vec{F}dt = \int_{p_i}^{p_f} d\vec{p} \rightarrow \vec{J} = \vec{F}\Delta t = \Delta\vec{p}$$

This equation relates impulse to change in momentum to force applied over a time interval. For the same change in momentum, force can vary by changing the time over which it is applied. Great examples include airbags in cars, boxers rolling with punches, skydivers bending their knees upon landing, etc. To summarize, when an unbalanced force acts on an object for a period of time, a change in momentum is produced, known as an impulse.

6.12 Q: A tow-truck applies a force of 2000N on a 2000-kg car for a period of 3 seconds.
(A) What is the magnitude of the change in the car's momentum?
(B) If the car starts at rest, what will be its speed after 3 seconds?

6.12 A: A) $\Delta p = F\Delta t = (2000N)(3s) = 6000N \cdot s$

B) $\Delta p = p - p_0 = mv - mv_0 \rightarrow v = \dfrac{\Delta p + mv_0}{m} = \dfrac{6000N \cdot s + 0}{2000kg} = 3\,m/s$

6.13 Q: A 2-kilogram body is initially traveling at a velocity of 40 meters per second east. If a constant force of 10 newtons due east is applied to the body for 5 seconds, the final speed of the body is
(A) 15 m/s
(B) 25 m/s
(C) 65 m/s
(D) 90 m/s
(E) 130 m/s

6.13 A: (C) $Ft = \Delta p = m\Delta v \rightarrow \Delta v = v - v_0 = \dfrac{Ft}{m} \rightarrow v = \dfrac{Ft}{m} + v_0 = \dfrac{(10N)(5s)}{2kg} + 40\,m\!/\!_s = 65\,m\!/\!_s$

6.14 Q: A motorcycle being driven on a dirt path hits a rock. Its 60-kilogram cyclist is projected over the handlebars at 20 meters per second into a haystack. If the cyclist is brought to rest in 0.50 seconds, the magnitude of the average force exerted on the cyclist by the haystack is
(A) 6.0×10^1 N
(B) 5.9×10^2 N
(C) 1.2×10^3 N
(D) 1.8×10^3 N
(E) 2.4×10^3 N

6.14 A: (E) $Ft = \Delta p = m\Delta v = m(v - v_0) \rightarrow F = \dfrac{m(v - v_0)}{t} = \dfrac{(60kg)(0 - 20\,m\!/\!_s)}{0.5s} = -2400N$

6.15 Q: The instant before a batter hits a 0.14-kilogram baseball, the velocity of the ball is 45 meters per second west. The instant after the batter hits the ball, the ball's velocity is 35 meters per second east. The bat and ball are in contact for 1.0×10^{-2} second. Calculate the magnitude of the average force the bat exerts on the ball while they are in contact.

6.15 A: $Ft = \Delta p = m\Delta v \rightarrow F = \dfrac{m\Delta v}{t} = \dfrac{m(v - v_0)}{t} = \dfrac{(0.14kg)(35\,m\!/\!_s - -45\,m\!/\!_s)}{1 \times 10^{-2}s} = 1120N$

6.16 Q: In an automobile collision, a 44-kilogram passenger moving at 15 meters per second is brought to rest by an air bag during a 0.10-second time interval. What is the magnitude of the average force exerted on the passenger during this time?
(A) 440 N
(B) 660 N
(C) 1320 N
(D) 4400 N
(E) 6600 N

6.16 A: (E) $Ft = \Delta p \rightarrow F = \dfrac{\Delta p}{t} = \dfrac{p - p_0}{t} = \dfrac{0 - (44kg)(15\,^m\!/_s)}{0.1s} = -6600N$

6.17 Q: The following carts are moving to the right across a frictionless surface with the specified initial velocity. A force is applied to each cart for a set amount of time as shown in the diagram.

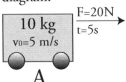

F=20N   t=5s

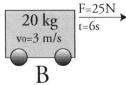

F=25N   t=6s

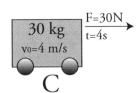

F=30N   t=4s

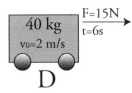

F=15N   t=6s

Rank the four carts from least to greatest in terms of

I) initial momentum   III) final momentum

II) impulse applied   IV) final velocity

6.17 A: I) A, B, D, C

II) D, A, C, B

III) A, D, B, C

IV) D, C, B, A

6.18 Q: A boy with a water gun shoots a stream of water that ejects 0.2 kg of water per second horizontally at a speed of 10 m/s. What horizontal force must the boy apply on the gun in order to hold it in position?

6.18 A: $F = \dfrac{dp}{dt} = \dfrac{d}{dt}(mv) = \dfrac{dm}{dt}v = \dfrac{0.2\,kg}{1\,s}(10\,^m\!/_s) = 2\,N$

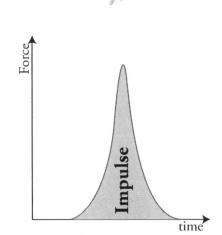

Impulse can also be found by taking the area under a force vs. time graph. The area under the graph is the change in momentum. For linear graphs, you can often times use geometry to find the area under the graph. For more complicated graphs, integration may be a better path to finding the area.

    Chapter 6: Momentum

6.19 Q: A time-varying force is applied to an object consistent with the force-time graph shown below. What is the magnitude of the impulse applied to the object?

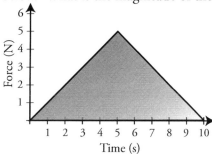

6.19 A: Determine the magnitude of the impulse by taking the area under the curve.
$$J = Area_\Delta = \tfrac{1}{2}bh = \tfrac{1}{2}(10s)(5N) = 25\ N \cdot s$$

6.20 Q: The graph indicates the force on a truck of mass 2000 kilograms as a function of time.

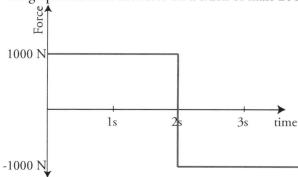

In the interval from 0 to 3 seconds, determine the change in the truck's velocity.

6.20 A: $\vec{J} = \displaystyle\int_{t=0s}^{3s} Fdt = (1000N)(2s) - 1000N(1s) = 2000N \cdot s - 1000N \cdot s = m\Delta v \rightarrow$
$$1000N \cdot s = (2000kg)\Delta v \rightarrow \Delta v = 0.5\,{}^m\!/_s$$

6.21 Q: A force $F(t)=t^3$ is applied to a 10-kilogram mass. What is the total impulse applied to the object between 1 and 3 seconds?

6.21 A: $F = \dfrac{dp}{dt} \rightarrow dp = Fdt \rightarrow dp = t^3 dt \rightarrow \displaystyle\int_{p_i}^{p_f} dp = \int_{t=1s}^{3s} t^3 dt \rightarrow p_f - p_i = \dfrac{t^4}{4}\Big|_{1s}^{3s} \rightarrow$
$$\Delta p = \dfrac{81}{4} - \dfrac{1}{4} \rightarrow \Delta p = 20\,{}^{kg \cdot m}\!/_s = J$$

Conservation of Linear Momentum

In an isolated system, where no external forces act, linear momentum is always conserved. Put more simply, in any closed system, the total momentum of the system remains constant. Therefore, an external force (an interaction with an outside object or system) is required to change the motion of an object's center of mass.

In the case of a collision or explosion (an event), if you add up the individual momentum vectors of all of the objects before the event, you'll find that they are equal to the sum of the momentum vectors of the objects after the event. Written mathematically, the law of conservation of momentum states:

$$\vec{p}_{initial} = \vec{p}_{final}$$

This is a direct outcome of Newton's 3rd Law.

In analyzing collisions and explosions, a momentum table can be a powerful tool for problem solving. To create a momentum table, follow these basic steps:
1. Identify all objects in the system. List them vertically down the left-hand column.
2. Determine the momenta of the objects before the event. Use variables for any unknowns.
3. Determine the momenta of the objects after the event. Use variables for any unknowns.
4. Add up all the momenta from before the event and set them equal to the momenta after the event.
5. Solve your resulting equation for any unknowns.

A **collision** is an event in which two or more objects approach and interact strongly for a brief period of time. Let's look at how the problem-solving strategy can be applied to a simple collision:

6.22 Q: A 2000-kg car traveling at 20 m/s collides with a 1000-kg car at rest at a stop sign. If the 2000-kg car has a velocity of 6.67 m/s after the collision, find the velocity of the 1000-kg car after the collision.

6.22 A: Call the 2000-kg car Car A, and the 1000-kg car Car B. You can then create a momentum table as shown below:

Objects	Momentum Before (kg·m/s)	Momentum After (kg·m/s)
Car A	2000×20=40,000	2000×6.67=13,340
Car B	1000×0=0	1000×v$_B$=1000v$_B$
Total	40,000	13,340+1000v$_B$

Because momentum is conserved in any closed system, the total momentum before the event must be equal to the total momentum after the event.

$$p_i = p_f \rightarrow 40,000 = 13,340 + 1000v_B \rightarrow v_B = \frac{40,000 - 13,340}{1000} = 26.7\,\text{m/s}$$

Not all problems are quite so simple, but problem solving steps remain consistent.

6.23 Q: On a snow-covered road, a car with a mass of 1.1×10^3 kilograms collides head-on with a van having a mass of 2.5×10^3 kilograms traveling at 8 meters per second. As a result of the collision, the vehicles lock together and immediately come to rest. Calculate the speed of the car immediately before the collision. [Neglect friction.]

6.23 A: Define the car's initial velocity as positive and the van's initial velocity as negative. After the collision, the two objects become one, therefore you can combine them in the momentum table.

Objects	Momentum Before (kg·m/s)	Momentum After (kg·m/s)
Car	$1100\times v_{car}=1100v_{car}$	0
Van	$2500\times-8=-20{,}000$	
Total	$-20{,}000+1100v_{car}$	0

$$p_i = p_f \rightarrow -20{,}000 + 1100v_{car} = 0 \rightarrow v_{car} = \frac{20{,}000}{1100} = 18.2\,^{m}\!/_{s}$$

6.24 Q: A 70-kilogram hockey player skating east on an ice rink is hit by a 0.1-kilogram hockey puck moving toward the west. The puck exerts a 50-newton force toward the west on the player. Determine the magnitude of the force that the player exerts on the puck during this collision.

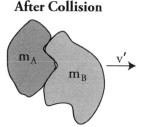

6.24 A: The player exerts a force of 50 newtons toward the east on the puck due to Newton's 3rd Law.

6.25 Q: The diagram below represents two masses before and after they collide.

Before Collision   **After Collision**

Before the collision, mass m_A is moving to the right with speed v, and mass m_B is at rest. Upon collision, the two masses stick together.

(A) $\dfrac{m_A + m_B v}{m_A}$ (C) $\dfrac{m_B v}{m_A + m_B}$

(B) $\dfrac{m_A + m_B}{m_A v}$ (D) $\dfrac{m_A v}{m_A + m_B}$

Which expression represents the speed, v', of the masses after the collision? (Assume no outside forces are acting on m_A or m_B.)

6.25 A: Use the momentum table to set up an equation utilizing conservation of momentum, then solve for the final velocity of the combined mass, labeled v'.

Objects	Momentum Before (kg·m/s)	Momentum After (kg·m/s)
Mass A	$m_A v$	$(m_A + m_B)v'$
Mass B	0	
Total	$m_A v$	$(m_A + m_B)v'$

(D) $m_A v = (m_A + m_B)v' \rightarrow v' = \dfrac{m_A v}{m_A + m_B}$

Let's take a look at another example which emphasizes the vector nature of momentum while examining an explosion. In physics terms, an **explosion** results when an object is broken up into two or more fragments.

6.26 Q: A 4-kilogram rifle fires a 20-gram bullet with a velocity of 300 m/s. Find the recoil velocity of the rifle.

6.26 A: Once again, you can use a momentum table to organize your problem-solving. To fill out the table, you must recognize that the initial momentum of the system is 0, and you can consider the rifle and bullet as a single system with a mass of 4.02 kg:

Objects	Momentum Before (kg·m/s)	Momentum After (kg·m/s)
Rifle	0	$4 \times v_{recoil}$
Bullet		$(.020)(300)=6$
Total	0	$6 + 4 \times v_{recoil}$

Due to conservation of momentum, you can again state that the total momentum before must equal the total momentum after, or $0=4v_{recoil}+6$. Solving for the recoil velocity of the rifle, you find:

$$p_i = p_f \rightarrow 0 = 4v_{recoil} + 6 \rightarrow v_{recoil} = \frac{-6}{4} = -1.5 \,^m\!/_s$$

The negative recoil velocity indicates the direction of the rifle's velocity. If the bullet traveled forward at 300 m/s, the rifle must travel in the opposite direction.

6.27 Q: The diagram below shows two carts that were initially at rest on a horizontal, frictionless surface being pushed apart when a compressed spring attached to one of the carts is released. Cart A has a mass of 3.0 kilograms and cart B has a mass of 5.0 kilograms.

If the speed of cart A is 0.33 meter per second after the spring is released, what is the approximate speed of cart B after the spring is released?
A) 0.12 m/s
B) 0.20 m/s
C) 0.33 m/s
D) 0.55 m/s

6.27 A: Define the positive direction toward the right of the page.

Objects	Momentum Before (kg·m/s)	Momentum After (kg·m/s)
Cart A	0	3×-0.33=-1
Cart B	0	5×v_B
Total	0	5v_B-1

(B) $p_i = p_f \rightarrow 0 = 5v_B - 1 \rightarrow v_B = \frac{1}{5} = 0.2 \,{}^m\!/_s$

6.28 Q: A woman jumps off a dock into a stationary boat with horizontal velocity v_1. After landing in the boat, the woman and the boat move with velocity v_2. Compared to velocity v_1, velocity v_2 has
A) the same magnitude and the same direction
B) the same magnitude and opposite direction
C) smaller magnitude and the same direction
D) larger magnitude and the same direction

6.28 A: (C) due to the law of conservation of momentum.

6.29 Q: A wooden block of mass m_1 sits on a floor attached to a spring in its equilibrium position. A bullet of mass m_2 is fired with velocity v into the wooden block, where it remains. Determine the maximum displacement of the spring if the floor is frictionless and the spring has spring constant k.

6.29 A: First, find the velocity of the bullet/block system after the bullet is embedded in the block.

Objects	Momentum Before (kg·m/s)	Momentum After (kg·m/s)
Block	0	$(m_1+m_2)v'$
Bullet	m_2v	
Total	m_2v	$(m_1+m_2)v'$

$$p_i = p_f \rightarrow m_2 v = (m_1 + m_2) v' \rightarrow v' = \frac{m_2 v}{m_1 + m_2}$$

Next, recognize the kinetic energy of the bullet-block system is completely converted to elastic potential energy in the spring.

$$K = U_S \rightarrow \tfrac{1}{2}(m_1 + m_2) v'^2 = \tfrac{1}{2} k x^2 \rightarrow x^2 = \frac{(m_1 + m_2) v'^2}{k} \quad \xrightarrow{v' = \frac{m_2 v}{m_1 + m_2}}$$

$$x^2 = \frac{(m_1 + m_2)}{k}\left(\frac{m_2 v}{m_1 + m_2}\right)^2 \rightarrow x = m_2 v \sqrt{\frac{1}{k(m_1 + m_2)}} = \frac{m_2 v}{\sqrt{k(m_1 + m_2)}}$$

Note: It may be tempting to begin by setting the initial kinetic energy of the bullet equal to the final elastic potential energy in the spring-bullet-block system and bypass analysis of the collision. This would be incorrect, however, as the collision is inelastic. A portion of the initial kinetic energy of the bullet is transferred into internal energy of the block, which is not transformed into elastic potential energy.

6.30 Q: Two carts of differing masses are held in place by a compressed spring on a frictionless surface. When the carts are released, allowing the spring to expand, which of the following quantities will have differing magnitudes for the two cars? (Choose all that apply.)
A) velocity
B) acceleration
C) force
D) momentum

6.30 A: (A) and (B) will have differing magnitudes. Each cart will experience the same magnitude of applied force due to Newton's 3rd Law, and the magnitude of the momentum of each cart will be the same due to the law of conservation of momentum.

6.31 Q: An open tub rolls across a frictionless surface. As it rolls across the surface, rain falls vertically into the tub. Which of the following statements best describe the behavior of the cart? (Choose all that apply.)
A) The tub will speed up.
B) The tub will slow down.
C) The tub's momentum will increase.
D) The tub's momentum will decrease.

6.31 A: (B) The tub will slow down. As the rain falls into the tub, the mass of the tub increases. Momentum must remain constant as no external force is applied; therefore, the velocity of the tub must decrease.

Types of Collisions

When objects collide, a number of different things can happen depending on the characteristics of the colliding objects. Of course, you know that momentum is always conserved in a closed system. Imagine, though, the differences in a collision if the two objects colliding are super bouncy balls compared to two lumps of clay. In the first case, the balls would bounce off each other. In the second, they would stick together and become, in essence, one object. Obviously, you need more ways to characterize collisions.

Elastic collisions occur when the colliding objects bounce off of each other. This typically occurs when you have colliding objects which are very hard or bouncy. Officially, an elastic collision is one in which the sum of the kinetic energy of all the colliding objects before the event is equal to the sum of the kinetic energy of all the objects after the event. Put more simply, kinetic energy is conserved in an elastic collision.

Inelastic collisions occur when two objects collide and kinetic energy is not conserved. In this type of collision some of the initial kinetic energy is converted into other types of energy (heat, sound, etc.), which is why kinetic energy is NOT conserved in an inelastic collision. In a perfectly inelastic collision, the two objects colliding stick together. In reality, most collisions fall somewhere between the extremes of a completely elastic collision and a completely inelastic collision.

6.32 Q: An astronaut floating in space is motionless. The astronaut throws her wrench in one direction, propelling her in the opposite direction. Which of the following statements are true? (Choose all that apply.)
A) The wrench will have a greater velocity than the astronaut.
B) The astronaut will have a greater momentum than the wrench.
C) The wrench will have greater kinetic energy than the astronaut.
D) The astronaut will have the same kinetic energy as the wrench.

6.32 A: (A) and (C) are both true. The wrench will have greater velocity than the astronaut because the astronaut and wrench will have the same magnitude of momentum due to the law of conservation of momentum, but the wrench has a smaller mass, so must have a larger velocity. The wrench will have greater kinetic energy than the astronaut due to the kinetic energy dependence on the square of velocity.

6.33 Q: Two billiard balls collide. Ball 1 moves with a velocity of 4 m/s, and ball 2 is at rest. After the collision, ball 1 comes to a complete stop. What is the velocity of ball 2 after the collision? Is this collision elastic or inelastic? The mass of each ball is 0.16 kg.

6.33 A: To find the velocity of ball 2, use a momentum table.

Objects	Momentum Before (kg·m/s)	Momentum After (kg·m/s)
Ball 1	0.16×4=0.64	0
Ball 2	0	0.16×v_2
Total	0.64	0.16×v_2

$p_i = p_f \rightarrow 0.64 = 0.16v_2 \rightarrow v_2 = 4 \, ^m/_s$

To determine whether this is an elastic or inelastic collision, you can calculate the total kinetic energy of the system both before and after the collision.

Objects	KE Before (J)	KE After (J)
Ball 1	0.5*0.16×4²=1.28	0
Ball 2	0	0.5*0.16×4²=1.28
Total	1.28	1.28

Since the kinetic energy before the collision is equal to the kinetic energy after the collision (kinetic energy is conserved), this is an elastic collision.

6.34 Q: A 0.5-kilogram tanker cart travels across a frictionless bathroom scale with an initial velocity of 10 cm/s. Beginning at time t=0, water is poured into the tank at a rate of 10 grams per second from a height of 20 cm.

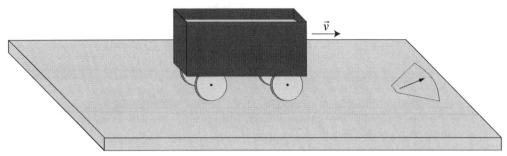

A) Determine an expression for the mass of the tanker cart as a function of time.

B) Derive an expression for the speed of the tanker cart as a function of time.

C) Determine the impulsive force required to stop the poured water in the tank.

D) Determine the reading on the scale as a function of time (assume the scale reads in units of newtons).

E) Is this an example of an elastic or inelastic collision? Explain.

6.34 A: A) Start by writing an equation for mass as a function of time, taking care to convert units to standard SI units to maintain consistency.

$$m(t) = m_{initial} + \frac{.01kg}{s}t = 0.5 + .01t$$

B) Beginning with the law of conservation of momentum, solve for the final velocity.

$$p_i = p_f \rightarrow m_i v_i = mv \rightarrow v = \frac{m_i v_i}{m} \xrightarrow[m=0.5+.01t]{m_i=0.5,\, v_i=0.1} v(t) = \frac{.05}{0.5+.01t}$$

C) Utilize the impulse-momentum theorem to recognize that the magnitude of the force of the water on the cart (and therefore the impulsive force required to stop the poured water by Newton's 3rd Law) is the rate of change of the momentum of the water. The velocity of the water as it hits the cart can be found using kinematics (neglecting any effects of the changing water level in the cart).

$$\Delta p_w = F_w \Delta t \rightarrow F_w = \frac{\Delta p_w}{\Delta t} = \frac{\Delta mv}{\Delta t} \xrightarrow[v_0=0,\, a=g,\, \Delta y=h]{v^2=v_0^2+2a\Delta y} F_w = \frac{\Delta m\sqrt{2gh}}{\Delta t} \xrightarrow[g=10,\, h=0.2]{\frac{\Delta m}{\Delta t}=0.01}$$

$$F_w = 0.01\sqrt{2(10)(0.2)} = 0.02N$$

D) Applying Newton's 2nd Law in the y-direction, the force of the scale upward (the normal force) is balanced by the weight of the cart and water as well as the force of the water being poured into the cart. From this expression, you can solve for the normal force as a function of time.

$$F_{net_y} = ma_y \xrightarrow{a_y=0} F_N = mg + F_{water} \xrightarrow[g=10,\, F_w=0.02]{m=0.5+.01t}$$

$$F_N = (0.5 + .01t)10 + 0.02 = 5.02 + 0.1t$$

E) Inelastic. Solve for the initial kinetic energy and compare to the kinetic energy as a function of time to quickly observe that the initial and final kinetic energies are not the same; therefore, kinetic energy is not conserved and the collision must be inelastic. $K_i \neq K_f$.

6.35 Q: A bullet of mass m_1 with velocity v_1 is fired into a block of mass m_2 attached by a string of length L in a device known as a ballistic pendulum.

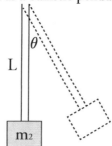

The ballistic pendulum records the maximum angle the string is displaced (θ). Determine the initial velocity of the bullet (v_1) in terms of m_1, m_2, L, g, and θ.

6.35 A: First determine the velocity of the bullet-block system after the collision.

Objects	Momentum Before (kg·m/s)	Momentum After (kg·m/s)
Bullet	m_1v	$(m_1+m_2)v'$
Block	0	
Total	m_1v	$(m_1+m_2)v'$

$$m_1v = (m_1 + m_2)v' \rightarrow v' = \frac{m_1}{m_1 + m_2}v$$

Next, recognize that the kinetic energy of the bullet-block system immediately after the collision is converted into gravitational potential energy at the highest point of the ballistic pendulum's swing. Use this relationship to solve for the initial velocity of the bullet.

$$K_{max} = \Delta U_g \rightarrow \tfrac{1}{2}mv'^2 = mgh \rightarrow v'^2 = 2gh \xrightarrow{v' = \left(\frac{m_1}{m_1+m_2}\right)v} \left(\frac{m_1}{m_1 + m_2}\right)^2 v^2 = 2gh \rightarrow$$

$$v^2 = \left(\frac{m_1}{m_1 + m_2}\right)^2 (2gh)$$

Finally, you need to place your answer in terms of L instead of h. This can be accomplished by recognizing the geometry of the ballistic pendulum in its initial and maximum displacement positions.

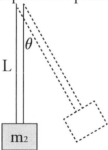 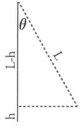

From here, an analysis of the triangle on the right allows you to solve for h in terms of L:

$$\cos\theta = \frac{adj}{hyp} \rightarrow \cos\theta = \frac{L - h}{L} \rightarrow h = L(1 - \cos\theta)$$

Finally, substitute your equation for the height of the block into your equation for the velocity of the bullet to provide the initial velocity of the bullet.

$$v^2 = \left(\frac{m_1 + m_2}{m_1}\right)^2 2gh \xrightarrow{h = L(1-\cos\theta)} v^2 = \left(\frac{m_1 + m_2}{m_1}\right)^2 2gL(1 - \cos\theta) \rightarrow$$

$$v = \left(\frac{m_1 + m_2}{m_1}\right)\sqrt{2gL(1 - \cos\theta)}$$

6.36 Q: Two carts of differing masses travel toward each other on a collision course as shown in the diagram below.

Cart 1   Cart 2

2 kg   v=2.5 m/s   v=1 m/s   1 kg

A) Determine the velocity of Cart 1 after the collision if Cart 2 moves to the right with a velocity of 2 m/s after the collision.

B) Is the collision elastic or inelastic?

C) If elastic, determine the kinetic energy of the system after the collision. If inelastic, identify at least one interaction that has not been considered that could account for the change in kinetic energy.

6.36 A: A) Use a momentum table to find the final velocity of Cart 1.

Objects	Momentum Before (kg·m/s)	Momentum After (kg·m/s)
Cart 1	2×2.5=5	2×v'
Cart 2	1×-1=-1	1×2
Total	4	2v'+2

Apply the law of conservation of momentum to solve for the final velocity of Cart 1.
$$p_i = p_f \rightarrow 4 = 2v' + 2 \rightarrow 2v' = 2 \rightarrow v' = 1\,{}^m\!/_s$$

B) Determine the total kinetic energy of the system before and after the collision.

Objects	K Before (J)	K After (J)
Cart 1	0.5*2×2.5²=6.25	0.5*2×1²=1
Cart 2	0.5*1×(-1)²=0.5	0.5*1×2²=2
Total	6.75	3

The kinetic energy before the collision is not equal to the kinetic energy after the collision; therefore this is an inelastic collision.

C) The change in kinetic energy could be accounted for by losses due to friction, deformation of the carts during the collision, sound, internal energy of the carts, etc.

6.37 Q: A traffic accident occurred in which a 3000-kg SUV traveling to the right rear-ended a 2000-kg car that was stopped at a stop sign. The diagram below depicts the skid marks of the scene after the accident.

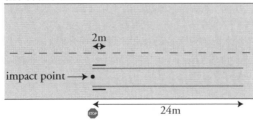

The acceleration of the car with the brakes locked is -3 m/s², and the acceleration of the SUV with the brakes locked is -2 m/s². Assuming both vehicles locked their brakes and began their skids at the instant of collision, determine the initial velocity of the truck.

6.37 A: Begin by determining the velocities of the two vehicles at the beginning of their skids (immediately after the collision) using kinematics.

Car:
$$v_x^2 = v_{x_0}^2 + 2a_x(x - x_0) \to v_{x_0}^2 = v_x^2 - 2a_x\Delta x \to v_{x_0}^2 = 0^2 - 2(-3)(24) \to$$
$$v_{x_0} = 12\,{}^m\!/_s$$

Truck: $v_{x_0}^2 = 0^2 - 2(-2)(2) \to v_{x_0} = 2.83\,{}^m\!/_s$

Next, create a momentum table using the velocities of the two vehicles immediately after the collision.

Objects	Momentum Before (kg·m/s)	Momentum After (kg·m/s)
Car	0	2000×12=24000
Truck	3000×v=3000v	3000×(2.83)=8490
Total	3000v	32,490

Finally, apply the law of conservation of momentum to find the velocity of the truck prior to the collision.
$$p_i = p_f \to 3000v = 32490 \to v = 10.83\,{}^m\!/_s$$

6.38 Q: A proton (mass=m) and a lithium nucleus (mass=7m) undergo an elastic collision as shown below.

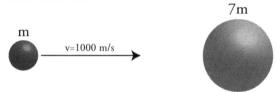

Find the velocity of the lithium nucleus following the collision.

6.38 A: First, solve for the velocity of the proton as a function of the velocity of the lithium nucleus using conservation of linear momentum.

Objects	Momentum Before (kg·m/s)	Momentum After (kg·m/s)
Proton	1000m	mv_p
Li Nucleus	0	$7mv_L$
Total	1000m	$mv_p + 7mv_L$

$$p_i = p_f \to 1000m = mv_p + 7mv_L \to 1000 = v_p + 7v_L \to v_p = 1000 - 7v_L$$

Next, utilize conservation of kinetic energy since this is an elastic collision to solve for the velocity of the lithium nucleus.
$$K_i = K_f \to \tfrac{1}{2}m(1000)^2 = \tfrac{1}{2}mv_p^2 + \tfrac{1}{2}(7m)v_L^2 \to 10^6 = v_p^2 + 7v_L^2 \to$$
$$v_p^2 = 10^6 - 7v_L^2 \xrightarrow{v_p = 1000 - 7v_L} (1000 - 7v_L)^2 = 10^6 - 7v_L^2 \to 56v_L^2 - 14000v_L = 0 \to$$
$$v_L(56v_L - 14000) = 0 \to v_L = 0 \ or \ v_L = 250\,{}^m\!/_s$$

From an analysis of the problem, choose the 250 m/s option.

Collisions in Multiple Dimensions

Much like the key to projectile motion (or two-dimensional kinematics problems) was breaking up vectors into their x- and y-components, the key to solving multi-dimensional collision problems involves breaking up momentum vectors into components parallel to the axes. The law of conservation of momentum then states that momentum is independently conserved in the x, y, and z directions.

$$p_{initial_x} = p_{final_x}$$

$$p_{initial_y} = p_{final_y}$$

$$p_{initial_z} = p_{final_z}$$

Therefore, you can solve multi-dimensional collision problems by creating separate momentum tables for the components of momentum before and after the event.

6.39 Q: Two objects of equal mass and velocities v_A and v_B collide as shown in the diagram below.

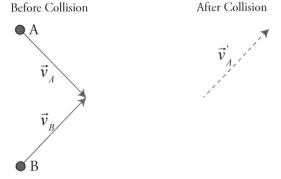

After the collision, object A travels with velocity v_A' as shown. Which vector best describes the velocity of object B after the collision?

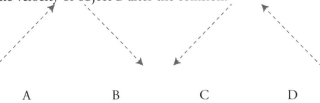

6.39 A: (B) is the only answer consistent with the law of conservation of momentum.

6.40 Q: Bert strikes a cue ball of mass 0.17 kg, giving it a velocity of 3 m/s in the x-direction. When the cue ball strikes the eight ball (mass=0.16 kg), previously at rest, the eight ball is deflected 45 degrees from the cue ball's previous path, and the cue ball is deflected 40 degrees in the opposite direction. Find the velocity of the cue ball and the eight ball after the collision.

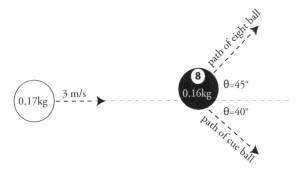

6.40 A: Start by making momentum tables for the collision, beginning with the x-direction. Since you don't know the velocity of the balls after the collision, call the velocity of the cue ball after the collision v_c, and the velocity of the eight ball after the collision v_8. Note that you must use trigonometry to determine the x-component of the momentum of each ball after the collision.

Objects	X-Momentum Before (kg·m/s)	X-Momentum After (kg·m/s)
Cue Ball	0.17×3=0.51	0.17×v_c×cos(-40°)
Eight Ball	0	0.16×v_8×cos(45°)
Total	0.51	0.17×v_c×cos(-40°)+ 0.16×v_8×cos(45°)

Since the total momentum in the x-direction before the collision must equal the total momentum in the x-direction after the collision, you can set the total before and total after columns equal:

$$p_i = p_f \rightarrow 0.51 = 0.17 v_c \cos\left(-40°\right) + 0.16 v_8 \cos\left(45°\right) = 0.130 v_c + 0.113 v_8$$

Next, create a momentum table and an algebraic equation for the conservation of momentum in the y-direction.

Objects	Y-Momentum Before (kg·m/s)	Y-Momentum After (kg·m/s)
Cue Ball	0	0.17×v_c×sin(-40°)
Eight Ball	0	0.16×v_8×sin(45°)
Total	0	0.17×v_c×sin(-40°)+ 0.16×v_8×sin(45°)

$$p_i = p_f \rightarrow 0 = 0.17 v_c \sin\left(-40°\right) + 0.16 v_8 \sin\left(45°\right) \rightarrow 0 = -0.109 v_c + 0.113 v_8$$

You now have two equations with two unknowns. To solve this system of equations, start by solving the y-momentum equation for v_c.

$$0 = -0.109 v_c + 0.113 v_8 \rightarrow v_c = 1.04 v_8$$

You can now take this equation for v_c and substitute it into the equation for conservation of momentum in the x-direction, effectively eliminating one of the unknowns, and giving a single equation with a single unknown.

$$p_i = p_f \rightarrow 0.51 = 0.130 v_c + 0.113 v_8 \xrightarrow{v_c = 1.04} 0.51 = 0.248 v_8 \rightarrow v_8 = 2.06\,{}^m\!/_s$$

Finally, solve for the velocity of the cue ball after the collision by substituting the known value for v_8 into the result of the y-momentum equation.

$$v_c = 1.04 v_8 \xrightarrow{v_8 = 2.06\,{}^m\!/_s} v_c = 1.04\left(2.06\,{}^m\!/_s\right) = 2.14\,{}^m\!/_s$$

6.41 Q: A cart traveling on a smooth track with velocity v collides and sticks to an identical cart on the track, initially at rest. What is the maximum percentage of the cart's initial kinetic energy maintained as kinetic energy in the two-cart system?

6.41 A: Consistent with the law of conservation of momentum, following the collision the maximum possible speed of the two carts combined is v/2. Kinetic energy, however, is related to the mass and the square of the velocity. Following the collision, the kinetic energy of the combined carts doubles due to the doubling of mass, but is quartered due to the effect of (v/2)². The product of these two effects, then, is an effective reduction in the kinetic energy of the combined carts by at least 50%. The missing energy must have been lost through non-conservative means such as heat and sound.

6.42 Q: A 0.5-kilogram block slides at 20 m/s on a smooth frictionless surface toward a stationary sphere, shown below left.

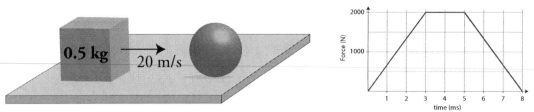

The sphere is half the volume of the block, but is eight times as dense. The block strikes the sphere at time t=0. A plot of the force exerted on the cube by the ball as a function of time is shown above right.

A) What is the impulse applied to the block?
B) What is the speed of the ball immediately following the collision?
C) What is the velocity of the cube immediately following the collision? (State both direction and magnitude.)
D) Is this an elastic collision? Justify your answer.

6.42 A: A) The impulse is the area under the curve which is 10 N•s.
B) The impulse applied is equal to the change of momentum, therefore

$$J = \Delta p = m\Delta v \rightarrow \Delta v = \frac{J}{m} = \frac{10N \cdot s}{2kg} = 5\,{}^m\!/_s$$

C) The change in velocity of the block can also be found from the impulse-momentum theorem:

$$J = \Delta p = m\Delta v \rightarrow \Delta v = \frac{J}{m} = \frac{-10N \cdot s}{0.5kg} = -20\,{}^m\!/_s$$

The velocity of the block after the collision is 0 m/s, since it was going 20m/s, and its change of velocity is -20 m/s. The direction is unnecessary—the block is stationary.
D) This is an inelastic collision since kinetic energy is not conserved. The initial kinetic energy of the block-sphere system is 100 joules, and the final kinetic energy of the block-sphere system is 25 joules.

6.43 Q: Two small, uniform balls of identical density and size are fired from a toy gun toward a wooden block. Ball A is highly elastic and bounces backward after striking the block. Ball B is made of clay and sticks to the wooden block upon impact.

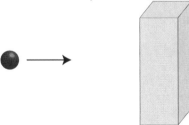

Which of the following statements best describes the effects of the collision with the block?
A) Ball A transfers more momentum and more energy to the block than Ball B.
B) Ball A transfers more momentum and less energy to the block than Ball B.
C) Ball A transfers less momentum and more energy to the block than Ball B.
D) Ball A transfers less momentum and less energy to the block than Ball B.

6.43 A: (B) Ball A transfers more momentum and less energy to the block than Ball B.

Both balls have the same initial momentum prior to striking the block. Following the collision, however, the elastic ball, Ball A, bounces backward, transferring up to twice its initial momentum to the block through the larger impulse. Ball B, however, sticks to the block, transferring its initial momentum to the block through an impulse equal to its initial momentum. Therefore, Ball A transfers more momentum to the block.

With respect to energy transfer, however, the story changes. Ball A maintains some kinetic energy as it rebounds off the block, and therefore cannot transfer as much kinetic energy to the block as Ball B, which transfers all of its kinetic energy to the block as it comes to rest. Therefore, Ball B transfers more kinetic energy to the block.

Center of Mass

The motion of real objects is considerably more complex than that of simple theoretical particles. As you model the motion of complex objects, you are really analyzing the motion of a system made up of collections of substructures, all in motion as pieces of a whole. You are able to do this by treating the system as if it were a point particle with its entire mass located at a specific location known as the object's **center of mass**.

For symmetric objects, it is typically easy to determine the center of mass of the object by inspection. For example, the center of mass of a uniform sphere is at the very center of the sphere. For more complex objects, such as an armadillo, more involved mathematical or empirical techniques are required to find the center of mass. Mathematically, the center of mass of an object is the weighted average of the location of mass in an object.

No matter how complex an object may be, you can find its center of mass and then treat the object as a point particle with total mass M. This allows you to apply basic physics principles to complex objects without adding unnecessary mathematical complexity to the analysis!

Center of Mass by Inspection

For uniform density and radially symmetric objects, the center of mass is the geometric center of the object. For objects with multiple parts, find the center of mass of each part and treat it as a point. You can find the center of mass for irregular objects by suspending the object from two or more points and dropping a plumb line. The plumb lines will intersect at the center of mass.

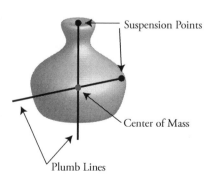

Suspension Points

Center of Mass

Plumb Lines

6.44 Q: Find the center of mass of the object shown below. The density of the object is uniform.

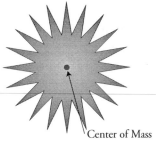

Center of Mass

6.44 A: The center of mass is in the geometric center, as shown.

Center of Mass for Systems of Particles

You can find the center of mass of a system of particles by taking the sum of the mass of the particles, multiplied by their positions, and dividing that by the total mass of the object. Looking at this in two dimensions, the center of mass in the x- and y-directions would be:

$$x_{cm} = \frac{\sum m_i x_i}{\sum m_i} = \frac{m_1 x_1 + m_2 x_2 + ...}{m_1 + m_2 + ...}, \quad y_{cm} = \frac{\sum m_i y_i}{\sum m_i} = \frac{m_1 y_1 + m_2 y_2 + ...}{m_1 + m_2 + ...}$$

6.45 Q: Find the center of mass of an object modeled as two separate masses on the x-axis. The first mass is 2 kg at an x-coordinate of 2m and the second mass is 6 kg at an x-coordinate of 8m.

6.45 A: $x_{cm} = \dfrac{\sum m_i x_i}{\sum m_i} = \dfrac{m_1 x_1 + m_2 x_2 + ...}{m_1 + m_2 + ...} = \dfrac{(2kg)\,(2m) + (6kg)\,(8m)}{(2kg + 6kg)} = 6.5m$

This means you can treat the object as a point particle with a mass of 8 kg at an x-coordinate of 6.5m as shown below. Note that we are performing all calculations from the perspective of the origin as the reference point, though you could use any reference point you prefer.

Use the same strategy for finding the center of mass of a multi-dimensional object.

6.46 Q: Find the coordinates of the center of mass for the system shown below.

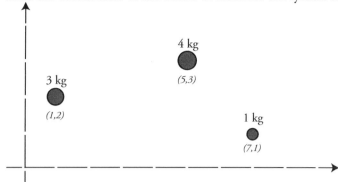

6.46 A:
$$x_{cm} = \frac{\sum m_i x_i}{\sum m_i} = \frac{(3kg)(1m) + (4kg)(5m) + (1kg)(7m)}{(3kg + 4kg + 1kg)} = 3.75m$$

$$y_{cm} = \frac{\sum m_i y_i}{\sum m_i} = \frac{(3kg)(2m) + (4kg)(3m) + (1kg)(1m)}{(3kg + 4kg + 1kg)} = 2.38m$$

Therefore the center of mass is a point particle with mass 8 kg located at (3.75m, 2.38m).

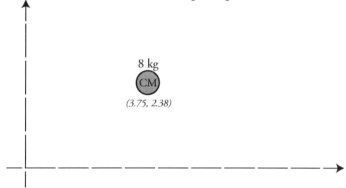

6.47 Q: A 30-kg raft of dimensions 3m x 3m is motionless on a lake. A 45 kg boy crosses from one corner of the raft to the other. Neglect friction.
A) How far does the center of mass of the raft/boy system move?
B) How far does the raft move?

6.47 A: A) There is no interaction outside the system (no external force), therefore the center of mass will not move.
B) First, find the center of mass of the system in its initial state. Assume the raft's mass is centered on the raft.

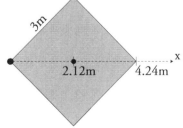

In setting up the problem, you can place the boy's path along the x-axis so that y-motion may be neglected in your analysis. Using this setup, and calling the left-most corner of the raft x=0, the boy initial starts at a position of x=0, and the raft can be modeled as a point particle of mass 30 kg at an x-position of 2.12m. The initial center of mass of the problem can then be found as:

$$x_{cm_i} = \frac{\sum m_i x_i}{\sum m_i} = \frac{(45kg)(0) + (30kg)(2.12m)}{(45kg + 30kg)} = 0.848m$$

Realizing the center of mass of the system cannot move without an external force, you can then solve for the new x-position of the raft.

$$x_{cm_f} = \frac{\sum m_i x_i}{\sum m_i} = \frac{(45kg)(4.24m) + (30kg)(x)}{(45kg + 30kg)} = 0.848m \rightarrow x = -4.24m$$

The raft started with its center at an x-position of 2.12m, and ended with its center at an x-position of -4.24m, which means it moved 6.36m in the direction opposite the boy's displacement.

6.48 Q: Find the center of mass of the combination object shown at right. The density of each shape is uniform.

6.48 A: By inspection of the symmetry of the object, the x-coordinate of the center of mass must be zero. Calculate the y-coordinate of the center of mass as follows:

$$y_{cm} = \frac{m_1 y_1 + m_2 y_2}{m_{total}} = \frac{(3kg)(0) + (6kg)(3m)}{9\ kg} = 2m$$

Therefore, the center of mass must be located at (0, 2m).

6.49 Q: An 80 kg student stands on the left end of a 240-kg log which is floating in the water (which you may treat as a frictionless surface). The student and the log are both initially at rest.

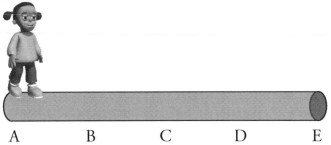

A   B   C   D   E

A) Which point is the approximate center of mass of the student-log system?
B) The student walks to the far end of the log at a constant speed of 2 m/s. As the student walks to the right, describe the motion of the log. Provide quantitative values wherever possible.
C) What is the velocity of the center of mass of the student-log system while the student is walking?

6.49 A: A) Location (B) is the approximate center of mass of the system.
B) The log moves to the left with a speed of 0.67 m/s as the student moves right.
C) The velocity of the center of mass of the student-log system is 0 m/s while the student is walking since the initial velocity of the center of mass of the system was 0 m/s and no external forces were applied.

Center of Mass by Integration

You can find the center of mass of more complex objects by summing up all the little pieces of position vectors multiplied by the differential of mass and then dividing by the total mass. The general formula for the position vector to the center of mass is:

$$\vec{r}_{CM} = \frac{\int \vec{r}\, dm}{M}$$

Let's demonstrate this by finding the center of mass of a uniform rod. Recognizing you could probably find the center of mass of this object by inspection, we'll use the integration method to demonstrate its utility.

6.50 Q: Find the center of mass of the uniform rod of length L and mass M shown below.

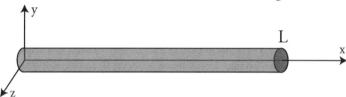

6.50 A: The mass of the rod is evenly distributed from x=0 to x=L. Define a **linear mass density** λ (lambda), which is the mass per unit length, λ =dm/dx.

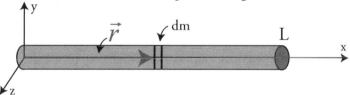

Because the rod is uniform, this linear mass density is a constant, which can be defined as λ =M/L. Use this relationship to find an expression for dm.

$$\lambda = \frac{dm}{dx} \rightarrow dm = \lambda dx$$

Now you can use the general formula for the center of mass.

$$\vec{r}_{CM} = \frac{\int \vec{r}\, dm}{M} \xrightarrow{\vec{r}=x} \vec{r}_{CM} = \frac{\int x\, dm}{M} \xrightarrow{dm=\lambda dx} \vec{r}_{CM} = \frac{\int x\lambda\, dx}{M}$$

The linear mass density λ (lambda) is a constant, so it can cross the integral sign, leaving you a straightforward calculation for the center of mass.

$$\vec{r}_{CM} = \frac{\int x\lambda dx}{M} = \frac{\lambda}{M}\int_{x=0}^{x=L} xdx = \frac{\lambda}{M}\frac{x^2}{2}\Big|_{x=0}^{x=L} = \frac{\lambda}{M}\left(\frac{L^2}{2} - \frac{0^2}{2}\right) = \frac{\lambda L^2}{2M}$$

Finally, recall the original definition of lambda as the linear mass density, equal to the total mass of the rod (M) divided by the length of the rod (L), which will allow you to simplify your result.

$$\vec{r}_{CM} = \frac{\lambda L^2}{2M} \xrightarrow{\lambda = \frac{M}{L}} \vec{r}_{CM} = \frac{ML^2}{2ML} = \frac{L}{2}$$

Therefore, as I'm sure you anticipated, the center of mass of the uniform rod is found at the center of the rod (half the length).

If you could have found the answer to this problem just by inspection, why bother with all that math and integration? The answer resides in the next problem. The derivation of the center of mass of a uniform rod is a great stepping stone to finding the center of mass of other more complex objects. Specifically, the AP-C exam states that students should be able to find the center of mass of a thin rod of non-uniform density. If you know how to find the center of mass of the uniform rod, the non-uniform rod is extremely similar!

6.51 Q: Find the center of mass of a non-uniform rod of length L and mass M with linear mass density given by the function $\lambda = kx$, as shown below.

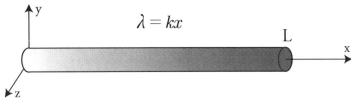

6.51 A: By inspection, we can estimate that the center of mass will be somewhere between $x=L/2$ and $x=L$, which will provide a nice check of the answer when the math is complete.

Start the formal derivation with the general relationship for the center of mass.

$$\vec{r}_{CM} = \frac{\int \vec{r} dm}{M} \xrightarrow{\vec{r}=x} \vec{r}_{CM} = \frac{\int x dm}{M}$$

What we'll actually do to calculate this is break up the rod into infinitessimally small pieces of width dx and mass dm. Then, we can integrate all the centers of mass of those little pieces from x=0 to x=L in order to get the entire rod's center of mass.

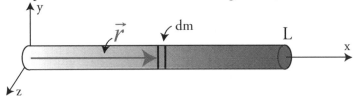

Next, using the relationship for the linear mass density, you can begin to solve for the center of mass.

$$\vec{r}_{CM} = \frac{\int x\,dm}{M} \xrightarrow[\lambda=\frac{dm}{dx}\to dm=(kx)\,dx]{\lambda=kx} \vec{r}_{CM} = \frac{\int x\,(kx)\,dx}{M}$$

Both k and M are constants, so they can cross the integral sign. With x as the variable of integration, you can integrate over the entire length of the non-uniform rod by starting at x=0 and finishing at x=L.

$$\vec{r}_{CM} = \frac{\int x\,(kx)\,dx}{M} \to \vec{r}_{CM} = \frac{k}{M}\int_{x=0}^{L} x^2\,dx = \frac{k}{M}\frac{x^3}{3}\Big|_{x=0}^{L} = \frac{k}{M}\left(\frac{L^3}{3} - \frac{0^3}{3}\right) = \frac{kL^3}{3M}$$

We now have an expression for the center of mass, but it is given in terms of both k and M, when what would be most helpful is an answer in terms of the length of the rod. You can develop an expression that relates M, the total mass of the rod, to the length of the rod, L. Derive this expression by integrating the linear mass density from x=0 to x=L for all the little pieces of rod dx.

$$M = \int_{x=0}^{x=L} \lambda\,dx \xrightarrow{\lambda=kx} M = \int_{x=0}^{x=L} kx\,dx = k\int_{x=0}^{x=L} x\,dx = k\frac{x^2}{2}\Big|_{x=0}^{L} = \frac{kL^2}{2}$$

You can then substitute this value for the total mass M into your expression for the center of mass of the non-uniform rod.

$$\vec{r}_{CM} = \frac{kL^3}{3M} \xrightarrow{M=\frac{kL^2}{2}} = \frac{kL^3}{3}\frac{2}{kL^2} = \frac{2L}{3}$$

Therefore, the center of mass of the non-uniform rod is located at two-thirds of its length, consistent with the estimate we determined before actually undertaking the problem.

Now that you can calculate center of mass, let's take a bit closer look at what it actually means. If you begin with our definition of the position vector and take the derivative of both sides with respect to time, you come up with an expression for the velocity of the center of mass as an infinite sum of infinitessimally small momenta.

$$\vec{r}_{CM} = \frac{1}{M}\sum_i m_i\vec{r}_i \xrightarrow{take\ derivatives} \vec{v}_{CM} = \frac{1}{M}\sum_i m_i\vec{v}_i \xrightarrow{\vec{p}=m\vec{v}} \vec{v}_{CM} = \frac{1}{M}\sum_i \vec{p}_i \to \sum_i \vec{p}_i = \vec{p}_{total} = M\vec{v}_{CM}$$

This indicates that the total momentum of the object or system can be found by multiplying the total mass by the velocity of the center of mass! You can take this another step further by taking the derivative of the resulting expression with respect to time.

$$\vec{p}_{total} = M\vec{v}_{CM} \xrightarrow{\vec{F}=\frac{d\vec{p}}{dt}} \frac{d\vec{p}_{total}}{dt} = \vec{F}_{total} = \frac{d}{dt}M\vec{v}_{CM} \to \frac{d\vec{p}_{total}}{dt} = \vec{F}_{total} = M\frac{d}{dt}\vec{v}_{CM} = M\vec{a}_{CM}$$

The resulting expression indicates that the total force on an object or system is equal to the total mass of the object or system multiplied by the acceleration of the center of mass. Again, you find that you can treat the complex object or system as a point particle located at the center of mass in this restatement of Newton's 2nd Law of Motion!

Similar to center of mass, **center of gravity** refers to the location at which the force of gravity acts upon an object as if it were a point particle with all its mass focused at that point. In a uniform gravitational field, center of mass and center of gravity are the same. In a non-uniform gravitational field, however, they may be different.

Chapter 7: Rotation

"Nothing good has ever been written about the full rotation of a race car about its roll axis."

—Carroll Smith

© Tatiana Shepeleva | stock.adobe.com

Objectives

1. Understand and apply relationships between translational and rotational kinematics.

2. Use the right hand rule to determine the direction of the angular velocity vector.

3. Determine the moment of inertia for combinations of point masses as well as continuous objects.

4. Apply the parallel axis theorem to find the moment of inertia of an object.

5. Calculate the torque on an object.

6. Apply Newton's 2nd Law for Rotation to find the angular acceleration of an object under the influence of a non-zero net torque.

7. Analyze systems involving strings and non-ideal pulleys.

8. Analyze systems involving objects that roll both with and without slipping.

9. Utilize the law of conservation of energy to analyze systems undergoing linear and rotational motion.

10. Determine the angular momentum of both particles and rotating objects.

11. Apply the law of conservation of angular momentum to analyze systems of revolving and rotating objects.

12. Utilize the relationship between net torque and angular momentum to analyze rotating bodies.

The motion of objects cannot always be described completely using the laws of physics that you've looked at so far. Besides motion, which changes an object's overall position (translational motion, or translational displacement), many objects also rotate around an axis, known as rotational, or angular, motion. The motion of some objects involves both translational and rotational motion.

An arrow speeding to its target, a hovercraft maneuvering through a swamp, and a hot air balloon floating through the atmosphere all experience only translational motion. A Ferris Wheel at an amusement park, a top spinning on a table, and a carousel at the beach experience only rotational motion. The Earth rotating around its axis (rotational motion) and moving through space (translational motion), and a frisbee spinning around its center while also flying through the air, both demonstrate simultaneous translational and rotational motion.

Rotational Kinematics

Rotational kinematics, exploring the motion of objects rotating about a point, is extremely similar to translational kinematics: all you have to do is learn the rotational versions of the kinematic variables and equations. When you learned translation kinematics, displacement was discussed in terms of $\Delta \vec{r}$, the change in position, and the distance traveled in terms of Δs. With rotational kinematics, you'll use the **angular displacement** vector $\vec{\theta}$ instead.

Looking at the example of an object traveling counterclockwise in a circular path, the linear position of the object is given by $\mathbf{r}$, and the distance traveled by s. The angular position vector is defined as θ, and angular displacement is $\Delta \theta$. You can then relate linear distance and angular displacement using the relation:

$$s = r\theta$$

When velocity was introduced in the translational world, you used the formula:

$$\vec{v} = \vec{r}\,' = \frac{d\vec{r}}{dt}$$

When exploring rotational motion, you'll talk about the **angular velocity** vector $\vec{\omega}$ (omega), given in units of radians per second (rad/s).

$$\vec{\omega} = \vec{\theta}\,' = \frac{d\vec{\theta}}{dt}$$

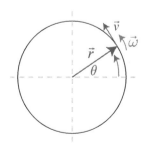

Because angular (rotational) velocity is a vector, define the positive direction of rotation as counter-clockwise around the circular path, and the negative direction as clockwise around the path. Looking at the diagram below, if you wrap the fingers of your right hand around the circular path in the direction the object movies, your thumb points in the direction of the angular velocity vector $\vec{\omega}$.

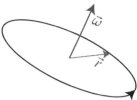

You can convert from linear speed to angular speed by first recognizing that linear speed is the derivative of distance traveled with respect to time.

$$v = \frac{ds}{dt}$$

Then, utilizing the relationship s=rθ, you can develop a simple translation:

$$v = \frac{ds}{dt} \xrightarrow{s=r\theta} v = \frac{dr\theta}{dt} \xrightarrow{r\,is\,a\,constant} v = r\frac{d\theta}{dt} \rightarrow v = r\omega$$

NOTE: Formally, the direction of angular vectors is determined by the right-hand rule. Wrap the fingers of your right hand in the direction of the rotational displacement, velocity, or acceleration, and your thumb will point in the vector's direction.

7.1 Q: A record spins on a phonograph at 33 rpms (revolutions per minute) clockwise. Find the angular velocity of the record.

7.1 A: $\omega = \dfrac{\Delta\theta}{\Delta t} = \dfrac{-33\ revs}{1\ min} = \dfrac{-33\ revs}{1\ min} \times \dfrac{1\ min}{60s} \times \dfrac{2\pi\ radians}{1\ rev} = -3.46\ ^{rad}/_s$

Note that the angular velocity vector is negative because the record is rotating in a clockwise direction.

7.2 Q: Find the magnitude of Earth's angular velocity in radians per second.

7.2 A: Realizing that the Earth makes one complete revolution every 24 hours, we can estimate the magnitude of the Earth's angular velocity as $\omega = \dfrac{\Delta\theta}{\Delta t} = \dfrac{2\pi\ radians}{24\ hours} \times \dfrac{1\ hour}{3600\ s} = 7.27 \times 10^{-5}\ ^{rad}/_s$

When acceleration was introduced in the translational regime, you used the formula:

$$\vec{a} = \vec{v}' = \frac{d\vec{v}}{dt}$$

The analog in rotational motion is the **angular acceleration** vector α (alpha), given in units of radians per second squared (rad/s²).

$$\vec{\alpha} = \vec{\omega}' = \frac{d\vec{\omega}}{dt}$$

Angular acceleration is the time rate of change of an object's angular velocity. The angular acceleration vector is positive for increasing angular velocities in the counter-clockwise direction, and negative for increasing angular velocities in the clockwise direction.

Following the same pattern that you used for velocity, you can develop a relationship between linera and angular acceleration:

$$a = \frac{dv}{dt} \xrightarrow{v = r\omega} a = \frac{dr\omega}{dt} \xrightarrow{r\ is\ a\ constant} a = r\frac{d\omega}{dt} \rightarrow a = r\alpha$$

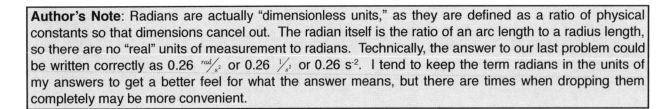

7.3 Q: "Dat Boi" the frog rides a unicycle. If the unicycle wheel begins at rest, and accelerates uniformly in a counter-clockwise direction to an angular velocity of 15 rpms in a time of 6 seconds, find the angular acceleration of the unicycle wheel.

7.3 A: First, convert 15 rpms to rads/s: $\frac{15\ revs}{1\ min} \times \frac{1\ min}{60\ s} \times \frac{2\pi\ radians}{1\ rev} = 1.57\ ^{rad}/_{s}$

Next, use the definition of angular acceleration:
$$\alpha = \frac{\Delta\omega}{\Delta t} = \frac{\omega - \omega_0}{t} = \frac{1.57\ ^{rad}/_{s} - 0}{6s} = 0.26\ ^{rad}/_{s^2}$$

Again, note the positive angular acceleration, as the bicycle wheel is accelerating in the counter-clockwise direction.

Author's Note: Radians are actually "dimensionless units," as they are defined as a ratio of physical constants so that dimensions cancel out. The radian itself is the ratio of an arc length to a radius length, so there are no "real" units of measurement to radians. Technically, the answer to our last problem could be written correctly as 0.26 $^{rad}/_{s^2}$ or 0.26 $^{1}/_{s^2}$ or 0.26 s⁻². I tend to keep the term radians in the units of my answers to get a better feel for what the answer means, but there are times when dropping them completely may be more convenient.

Putting this all together, you can find rotational (angular) parallels to the standard translational kinematic variables.

Variable	Translational	Angular
Distance	Δs	$\Delta \theta$
Velocity	v	ω
Acceleration	a	α
Time	t	t

You can convert from translational to angular quantities in a straightforward fashion.

Variable	Translational	Angular
Distance	$s = r\theta$	$\theta = \dfrac{s}{r}$
Velocity	$v = r\omega$	$\omega = \dfrac{v}{r}$
Acceleration	$a = r\alpha$	$\alpha = \dfrac{a}{r}$

Further, just as you had a set of kinematic equations to assist in solving translational motion problems dealing with constant acceleration, you can develop a set of rotational kinematic equations applicable to situations involving constant angular acceleration.

Translational	Rotational
$v = v_0 + at$	$\omega = \omega_0 + \alpha t$
$\Delta x = v_0 t + \frac{1}{2}at^2$	$\Delta\theta = \omega_0 t + \frac{1}{2}\alpha t^2$
$v^2 = v_0^2 + 2a\Delta x$	$\omega^2 = \omega_0^2 + 2\alpha\Delta\theta$

An interesting exercise in utilizing these relationships can be found in deriving the centripetal acceleration formula. Begin by expressing the position vector **r** in terms of unit vectors for an object traveling in a counter-clockwise circular path as indicated in the diagram at right.

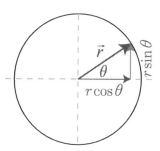

$$\vec{r} = r\cos(\theta)\hat{i} + r\sin(\theta)\hat{j} \xrightarrow{\theta = \omega t} \vec{r} = r\cos(\omega t)\hat{i} + r\sin(\omega t)\hat{j}$$

Taking the derivative of both sides allows you to develop an expression for the velocity of the object in unit vector notation.

$$\vec{v} = \frac{d\vec{r}}{dt} = \frac{d}{dt}(r\cos(\omega t)\hat{i} + r\sin(\omega t)\hat{j}) = \frac{d}{dt}(r\cos(\omega t)\hat{i}) + \frac{d}{dt}(r\sin(\omega t)\hat{j}) \to$$

$$\vec{v} = r\hat{i}\frac{d}{dt}(\cos\omega t) + r\hat{j}\frac{d}{dt}(\sin\omega t) = r\hat{i}\omega(-\sin(\omega t)) + r\hat{j}\omega(\cos(\omega t)) = -\omega r\sin(\omega t)\hat{i} + \omega r\cos(\omega t)\hat{j}$$

Next, take the derivative of the velocity with respect to time to find the acceleration.

$$\vec{a} = \frac{d\vec{v}}{dt} = \frac{d}{dt}(-\omega r\sin(\omega t)\hat{i} + \omega r\cos(\omega t)\hat{j}) = -\omega^2 r\cos(\omega t)\hat{i} + -\omega^2 r\sin(\omega t)\hat{j} \to$$

$$\vec{a} = -\omega^2(r\cos(\omega t)\hat{i} + r\sin(\omega t)\hat{j})$$

Notice that the resulting expression includes the original expression for the position function **r**.

$$\vec{a} = -\omega^2(r\cos(\omega t)\hat{i} + r\sin(\omega t)\hat{j}) \xrightarrow{\vec{r} = r\cos(\omega t)\hat{i} + r\sin(\omega t)\hat{j}} \vec{a} = -\omega^2 r \xrightarrow{\omega = \frac{v}{r}} a = -\left(\frac{v}{r}\right)^2 r$$

Finally, note that we originally defined the position function **r** starting at the center and moving outward. If instead we define the centripetal acceleration vector as positive toward the center of the circle as is traditional, you obtain the standard formula for centripetal acceleration.

$$a_c = \left(\frac{v}{r}\right)^2 r = \frac{v^2}{r}$$

7.4 Q: An object of mass m moves in a circular path of radius r=2m according to $\theta = 2t^3 + t + 4$, where θ is measured in radians and t is in seconds.
A) Determine the angular velocity of the object at time t=2 seconds.
B) Determine the object's speed at this time.

7.4 A: A) $\omega = \dfrac{d\theta}{dt} = \dfrac{d}{dt}\left(2t^3 + t + 4\right) = 6t^2 + 1 \xrightarrow{t=2s} \omega = 6\left(2^2\right) + 1 = 25\,{}^{rad}\!/_s$

B) $v\left(t = 2s\right) = r\omega = \left(2m\right)\left(25\,{}^{rad}\!/_s\right) = 50\,{}^m\!/_s$

7.5 Q: A wheel of radius r and mass M undergoes a constant angular acceleration of magnitude α. What is the speed of the wheel after it has completed one full rotation, assuming it started from rest?

7.5 A: $\omega^2 = \omega_0^2 + 2\alpha\Delta\theta \xrightarrow[\Delta\theta = 2\pi]{\omega_0 = 0} \omega^2 = 0^2 + 2\alpha\left(2\pi\right) = 4\pi\alpha \rightarrow \omega = \sqrt{4\pi\alpha} \xrightarrow{v = r\omega} v = r\sqrt{4\pi\alpha}$

7.6 Q: A uniform hollow tube of length L rotates vertically about its center of mass as shown. A ball is dropped into the tube at position A, and exits a short time later at position B. From the perspective of a stationary observer watching the tube rotate, the distance the ball travels is
(A) less than L
(B) greater than L
(C) equal to L

7.6 A: (B) greater than L. Though the displacement of the ball at B from its initial position at A is less than the length of the rod, L, the distance the ball travels is greater than L from the point of view of an external observer watching the tube rotate. Imagine the tube is transparent as you observe the path of the ball from a stationary reference point. You would observe the ball traveling a curved path from A to B, as shown at right. The length of A to the center point is one radius, and since the ball takes a curved path, it travels a distance greater than that radius. The same occurs from the center point to point B. Therefore, the ball travels a distance greater than the length of the tube from the perspective of an external observer at a stationary reverence point.

7.7 Q: A carpenter cuts a piece of wood with a high powered circular saw. The saw blade accelerates from rest with an angular acceleration of 14 rad/s² to a maximum speed of 15,000 rpms. What is the maximum speed of the saw in radians per second?

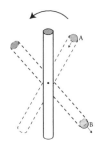

7.7 A: $\dfrac{15000\ revs}{1\ min} \times \dfrac{1\ min}{60\ s} \times \dfrac{2\pi\ radians}{1\ rev} = 1570\ {}^{rad}\!/_s$

7.8 Q: How long does it take the saw to reach its maximum speed?

7.8 A: You can use the rotational kinematic equations to solve this problem:

Variable	Value
ω_0	0 rad/s
ω	1570 rad/s
$\Delta\theta$	?
α	14 rad/s²
t	FIND

$$\omega = \omega_0 + \alpha t \rightarrow t = \frac{\omega - \omega_0}{\alpha} = \frac{1570\,^{rad}\!/_s - 0}{14\,^{rad}\!/_{s^2}} = 112\,s$$

7.9 Q: How many complete rotations does the saw make while accelerating to its maximum speed?

7.9 A: $\theta = \theta_0 + \omega_0 t + \frac{1}{2}\alpha t^2 = \frac{1}{2}(14\,^{rad}\!/_{s^2})(112s)^2 = 87,800\,rads$

$$87,800\,rads \times \frac{1\,rev}{2\pi\,rads} = 14,000\;revolutions$$

7.10 Q: A safety mechanism will bring the saw blade to rest in 0.3 seconds should the carpenter's hand come off the saw controls. What angular acceleration does this require? How many complete revolutions will the saw blade make in this time?

7.10 A: Begin by re-creating the rotational kinematics table.

Variable	Value
ω_0	1570 rad/s
ω	0 rad/s
$\Delta\theta$	FIND
α	FIND
t	0.3s

First, find the angular acceleration.
$$\omega = \omega_0 + \alpha t \rightarrow \alpha = \frac{\omega - \omega_0}{t} = \frac{0 - 1570\,^{rad}\!/_s}{0.3s} = -5230\,^{rad}\!/_{s^2}$$

Next, find the angular displacement.
$$\Delta\theta = \omega_0 t + \frac{1}{2}\alpha t^2 = (1570\,^{rad}\!/_s)(0.3s) + \frac{1}{2}(-5230\,^{rad}\!/_{s^2})(0.3s)^2 = 236\;radians$$

Finally, convert the angular displacement into revolutions
$$\Delta\theta = 236\,rads \times \frac{1\,rev}{2\pi\,rads} = 37.5\;revolutions$$

7.11 Q: An amusement park ride of radius x allows children to sit in a spinning swing held by a cable of length L.

At maximum angular speed, the cable makes an angle of θ with the vertical as shown in the diagram below. Determine the maximum angular speed of the rider in terms of g, θ, x and L.

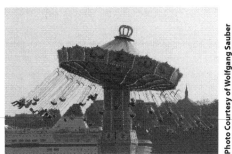

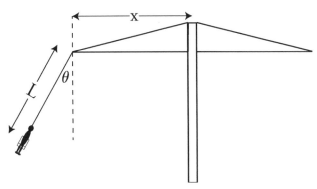

7.11 A: First write a Newton's 2nd Law equation for the x-direction, calling the tension in the cable T. Note that since the rider is moving in a circular path, the net force in the x-direction must also be the rider's centripetal force.

$$F_{net_x} = T\sin\theta = ma = \frac{mv^2}{r}$$

Then, write a Newton's 2nd Law equation for the y-direction.
$$F_{net_y} = T\cos\theta - mg = 0 \rightarrow T\cos\theta = mg$$

Now notice that you can divide the first equation by the second equation to eliminate T.
$$\frac{T\sin\theta}{T\cos\theta} = \frac{\frac{mv^2}{r}}{mg} = \frac{mv^2}{mgr} \rightarrow \tan\theta = \frac{v^2}{gr}$$

Next, convert translational speed to angular speed and solve for angular speed ω.
$$\tan\theta = \frac{v^2}{gr} \xrightarrow{v=\omega r} \tan\theta = \frac{\omega^2 r}{g} \rightarrow \omega = \sqrt{\frac{g\tan\theta}{r}}$$

Finally, format your answer in terms of g, θ, x and L by replacing r with x+Lsinθ.
$$\omega = \sqrt{\frac{g\tan\theta}{r}} \xrightarrow{r=x+L\sin\theta} \omega = \sqrt{\frac{g\tan\theta}{x+L\sin\theta}}$$

Frequency and Period

For objects moving in circular paths, you can characterize their motion around the circle using the terms frequency (f) and period (T). The **frequency** of an object is the number of revolutions the object makes in a complete second. It is measured in units of [1/s], or Hertz (Hz). In similar fashion, the **period** of an object is the time it takes to make one complete revolution. Since the period is a time interval, it is measured in units of seconds. You can relate period and frequency using the equations:

$$f = \frac{1}{T} \qquad T = \frac{1}{f}$$

7.12 Q: A 500g toy train completes 10 laps of its circular track in 1 minute and 40 seconds. If the diameter of the track is 1 meter, find the train's centripetal acceleration (a_c), centripetal force (F_c), period (T), and frequency (f).

7.12 A: $v = \dfrac{d}{t} = \dfrac{10\,(2\pi r)}{t} = \dfrac{(10)\,2\pi\,(0.5m)}{100s} = 0.314 \,{}^{m}\!/_{s}$

$a_c = \dfrac{v^2}{r} = \dfrac{(0.314 \,{}^{m}\!/_{s})^2}{0.5m} = 0.197 \,{}^{m}\!/_{s^2}$

$F_c = ma_c = (0.5kg)\,(0.197 \,{}^{m}\!/_{s^2}) = 0.099N$

$T = \dfrac{100s}{10\ revs} = 10s$

$f = \dfrac{1}{T} = \dfrac{1}{10s} = 0.1\ Hz$

7.13 Q: Alan makes 38 complete revolutions on the playground Round-A-Bout in 30 seconds. If the radius of the Round-A-Bout is 1 meter, determine
(A) Period of the motion
(B) Frequency of the motion
(C) Speed at which Alan revolves
(D) Centripetal force on 40-kg Alan

7.13 A: (A) $T = \dfrac{30s}{38\ revs} = 0.789s$

(B) $f = \dfrac{1}{T} = \dfrac{1}{0.789s} = 1.27\ Hz$

(C) $v = \dfrac{d}{t} = \dfrac{38 \times 2\pi r}{t} = \dfrac{38 \times 2\pi\,(1m)}{30s} = 7.96 \,{}^{m}\!/_{s}$

(D) $F_c = ma_c = m\dfrac{v^2}{r} = (40kg)\dfrac{(7.96 \,{}^{m}\!/_{s})^2}{1m} = 2530N$

Moment of Inertia

Previously, the inertial mass of an object (its translational inertia) was defined as that object's ability to resist a linear acceleration. The symbol for an object's inertial mass is m. Similarly, an object's **rotational inertia**, or **moment of inertia**, describes an object's resistance to a rotational (or angular) acceleration. The symbol for an object's moment of inertia is I.

Objects that have most of their mass near their axis of rotation have smaller moments of inertia, while objects that have more mass farther from the axis of rotation have larger moments of inertia. Consider the case of a toy top compared to a carousel. Although both will rotate, it takes considerably less effort to rotate a top at a specific angular velocity than it does to rotate the carousel at the same angular velocity. The carousel has a significantly larger moment of inertia, or rotational inertia, than the top.

For common objects, you can look up the formula for their moment of inertia. For more complex objects, the moment of inertia can be calculated by taking the sum of all the individual particles of mass making up the object multiplied by the square of their radius from the axis of rotation.

$$I = \sum mr^2 = \int r^2 dm$$

Commonly Used Moments of Inertia

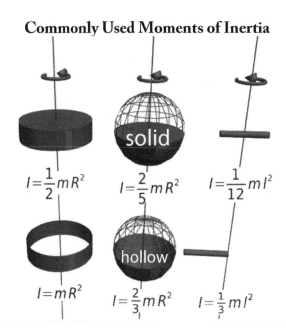

$$I = \frac{1}{2}mR^2 \qquad I = \frac{2}{5}mR^2 \qquad I = \frac{1}{12}ml^2$$

$$I = mR^2 \qquad I = \frac{2}{3}mR^2 \qquad I = \frac{1}{3}ml^2$$

7.14 Q: Calculate the moment of inertia for a solid sphere with a mass of 10 kg and a radius of 0.2m.

7.14 A: Utilize the Commonly Used Moments of Inertia to find:
$$I = \tfrac{2}{5}mR^2 = \tfrac{2}{5}(10kg)(0.2m)^2 = 0.16 kg \cdot m^2$$

7.15 Q: Find the moment of inertia (I) of two 5-kg bowling balls joined by a meter-long rod of negligible mass when rotated about the center of the rod. Compare this to the moment of inertia of the object when rotated about one of the masses.

7.15 A: $I = \sum mr^2 = (5kg)(0.5m)^2 + (5kg)(0.5m)^2 = 2.5kg \cdot m^2$

$I = \sum mr^2 = (5kg)(1m)^2 + (5kg)(0m)^2 = 5kg \cdot m^2$

7.16 Q: Calculate the moment of inertia for a long thin rod with a mass of 2 kg and a length of 1 meter rotating around the center of its length.

7.16 A: $I = \frac{1}{12}ml^2 = \frac{1}{12}(2kg)(1m)^2 = 0.17kg \cdot m^2$

7.17 Q: Calculate the moment of inertia for a long thin road with a mass of 2 kg and a length of 1 meter rotating about its end.

7.17 A: $I = \frac{1}{3}ml^2 = \frac{1}{3}(2kg)(1m)^2 = 0.67kg \cdot m^2$

You can derive the moment of inertia of various objects from the definition of moment of inertia. Let's start with the moment of inertia of a uniform rod about its end and about its center, and compare to the known solutions from the Commonly Used Moments of Inertia table.

7.18 Q: Find the moment of inertia of a uniform rod about its end, and compare that to the moment of inertia of the rod about its center.

 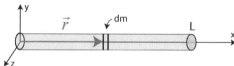

7.18 A: Begin by defining a linear mass density, λ, equal to the total mass M divided by the length of the rod L.

$$\lambda = \frac{M}{L}$$

If you then slide up the rod into infinitessimal thin sections, as shown above right, the mass contained in each infinitessimally small section is the linear mass density λ multiplied by the width of the infinitessimally small section, dx.

$dm = \lambda dx$

You may now solve for the moment of inertia of the rod rotating about its end using the definition of the moment of inertia.

$$I = \int r^2 dm \xrightarrow[r=x]{dm=\lambda dx} I = \int_{x=0}^{x=L} x^2 \lambda dx \rightarrow I = \lambda \int_0^L x^2 dx = \lambda \frac{x^3}{3}\Big|_0^L = \lambda \frac{L^3}{3} \xrightarrow{\lambda = \frac{M}{L}}$$

$$I = \frac{M}{L}\frac{L^3}{3} = \frac{1}{3}ML^2$$

Note how the limits of integration are used to begin the integration at the pivot point (x=0) and finish at the end of the rod (x=L). As expected, the derived solution matches the given solution in the table. We can follow the same strategy, but with adjusted limits of integration, to derive the moment of inertia of the rod about its center point.

$$I = \int r^2 \, dm \xrightarrow{dm=\lambda dx} I = \int_{x=-\frac{L}{2}}^{x=\frac{L}{2}} x^2 \lambda \, dx = \lambda \int_{x=-\frac{L}{2}}^{x=\frac{L}{2}} x^2 \, dx = \lambda \frac{x^3}{3} \Big|_{-\frac{L}{2}}^{\frac{L}{2}} = \frac{\lambda}{3}\left(\frac{L^3}{8} - \frac{-L^3}{8}\right) \rightarrow$$

$$I = \frac{\lambda}{3}\frac{L^3}{4} \xrightarrow{\lambda=\frac{M}{L}} I = \frac{M}{3L}\frac{L^3}{4} = \tfrac{1}{12}ML^2$$

As anticipated, the moment of inertia is considerably larger when the rod is rotated about its end, indicating it is harder to accelerate the rod in a circular path in that circumstance as more of the mass is further from the pivot point.

7.19 Q: Find the moment of inertia of a uniform solid cylinder about its axis.

7.19 A: First define a mass volume density, ρ, as the total mass divided by the volume of the cylinder.

$$\rho = \frac{M}{V} = \frac{M}{\pi R^2 L}$$

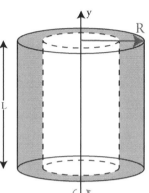

Strategy for the derivation will be to dissect the cylinder into concentric infinitessimally thin hollow cylinders, each with mass dm. You can then integrate from the center (r=0) to the outer edge of the cylinder (r=R) to add up all the moments of the thin hollow cylinders to obtain the moment of inertia of the entire solid cylinder.

The differential mass in the thin hollow cylinders can be found by finding the volume of the hollow cylinder and multiplying by the mass volume density. Imagine taking the thin hollow cylinder, cutting it along an edge, and laying it out flat. The shape would make a thin sheet with length L, width equal to the circumference ($2\pi r$), and depth dr.

$$I = \int r^2 \, dm \xrightarrow{dm=2\pi r\rho L dr} I = \int_{r=0}^{r=R} r^2 (2\pi r\rho L dr) = 2\pi\rho L \int_{r=0}^{R} r^3 \, dr = 2\pi\rho L \frac{R^4}{4} \xrightarrow{\rho=\frac{M}{\pi R^2 L}}$$

$$I = 2\pi \frac{M}{\pi R^2 L} L \frac{R^4}{4} \rightarrow I = \tfrac{1}{2}MR^2$$

Parallel Axis Theorem

If you know the moment of inertia of any object through an axis intersecting the center of mass of the object, you can find the moment of inertia around any parallel axis using the **Parallel Axis Theorem**. In the diagram below right, assume you know the moment of inertia of the object through axis l, which passes through the object's center of mass. The object has a mass m. You can find the moment of inertia of the object through axis l', which is parallel to l and located a distance d away from l, as follows.

$$I_{l'} = I_l + md^2$$

7.20 Q: Given the moment of inertia of a rod around its center ($I = \frac{1}{12}ML^2$), use the Parallel Axis Theorem to find the moment of inertia of the rod about its end.

7.20 A:
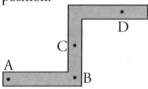

$$I_{end} = I_{center} + md^2 \xrightarrow[d=L/2]{I_{center}=\frac{1}{12}ML^2} I_{end} = \frac{1}{12}ML^2 + M\left(\frac{L}{2}\right)^2 = \frac{1}{3}ML^2$$

7.21 Q: An object with uniform mass density is rotated about an axle, which may be in position A, B, C, or D. Rank the object's moment of inertia from smallest to largest based on axle position.

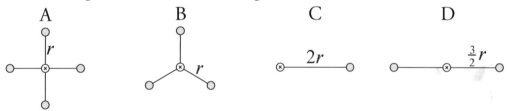

7.21 A: C, B, D, A

7.22 Q: Identical point masses are arranged in space and connected by massless rods in four different configurations, as shown in the diagram below.

Rank the moment of inertia of the configurations from greatest to least, assuming the masses are rotated about the point indicated with an x.

7.22 A: D>A=C>B. Moment of inertia can be obtained by taking the sum of the masses times the square of their distance from the axis of rotation.

7.23 Q: A uniform rod of length L has moment of inertia I_0 when rotated about its midpoint, shown below left. A sphere of mass M is added to each end of the rod, shown below right.

What is the new moment of inertia of the rod/ball system?

7.23 A: $I = I_0 + \sum mr^2 = I_0 + M\left(\frac{L}{2}\right)^2 + M\left(\frac{L}{2}\right)^2 = I_0 + \frac{ML^2}{4} + \frac{ML^2}{4} = I_0 + \frac{ML^2}{2}$

Torque

Torque (τ) is a force that causes an object to turn. It's the rotational version of force. If you think about using a wrench to tighten a bolt, the closer to the bolt you apply the force, the harder it is to turn the wrench, while the farther from the bolt you apply the force, the easier it is to turn the wrench. This is because you generate a larger torque when you apply a force at a greater distance from the axis of rotation.

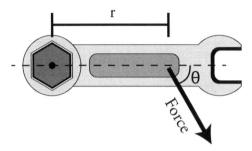

Let's take a look at the example of a wrench turning a bolt. A force is applied at a distance from the axis of rotation. The vector from the axis of rotation to the point at which the force is applied is vector r. When you apply forces at 90 degrees to the imaginary line leading from the axis of rotation to the point where the force is applied, you obtain maximum torque. As the angle at which the force applied decreases (θ), so does the torque causing the bolt to turn. Torque can be calculated as the cross product of the r and F vectors.

$$\vec{\tau} = \vec{r} \times \vec{F}$$

From the definition of the cross product, you can find the magnitude of the torque as follows:

$$|\vec{\tau}| = |\vec{F}||\vec{r}|\sin\theta = r_\perp F = rF\sin\theta$$

The direction of the torque vector can be determined from a right-hand rule. Point the fingers of your right hand in the direction of the **r** vector, and bend your fingers inward in the direction of the force vector. Your thumb will then point in the direction of the torque vector, which is perpendicular to both **r** and **F**. Positive torques cause counter-clockwise angular accelerations, and negative torques cause clockwise angular accelerations.

In some cases, physicists will refer to $r\sin\theta$ (or $r_\perp$) as the **lever arm**, or **moment arm**, of the system. The lever arm is the perpendicular distance from the axis of rotation to the point where the force is applied. The further away from the axis of rotation the force is applied the greater the lever arm, the greater the "leverage." Alternately, you could think of torque as the component of the force perpendicular to the lever multiplied by the distance r. Units of torque are the units of force × distance, or newton-meters (N·m).

Objects which have no rotational acceleration, or a net torque of zero, are said be in **rotational equilibrium**. This implies that any net positive (counter-clockwise) torque is balanced by an equal net negative (clockwise) torque, and the object will remain in its current state of rotation.

7.24 Q: A pirate captain takes the helm and turns the wheel of his ship by applying a force of 20 Newtons to a wheel spoke. If he applies the force at a radius of 0.2 meters from the axis of rotation, at an angle of 80° to **r**, what is the magnitude of the torque applied to the wheel?

7.24 A: $|\vec{\tau}| = |\vec{r}\,\|\,\vec{F}|\sin\theta = (0.2m)(20N)\sin(80°) = 3.94\ N \cdot m$

7.25 Q: A mechanic tightens the lugs on a tire by applying a torque of 110 N·m at an angle of 90° to the line of action.
A) What force is applied if the wrench is 0.4 meters long?
B) What is the minimum length of the wrench if the mechanic is only capable of applying a force of 200N?

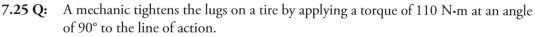

7.25 A: A) $|\vec{\tau}| = |\vec{r}\,\|\,\vec{F}|\sin\theta \longrightarrow F = \dfrac{\tau}{r\sin\theta} = \dfrac{110N \cdot m}{(0.4m)\,sin90°} = 275\ N$

B) $|\vec{\tau}| = |\vec{r}\,\|\,\vec{F}|\sin\theta \longrightarrow \tau = \dfrac{r}{F\sin\theta} = \dfrac{110N \cdot m}{(200N)\sin90°} = 0.55\ m$

7.26 Q: A constant force F is applied for five seconds at various points to the object below, as shown in the diagram. Rank the magnitude of the torque exerted by the force on the object about an axle located at the center of mass from smallest to largest.

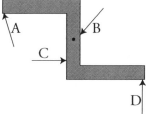

7.26 A: B, C, A, D

7.27 Q: A variety of masses are attached at different points to a uniform beam attached to a pivot. Rank the angular acceleration of the beam from largest to smallest.

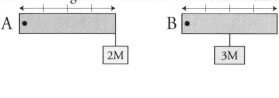

7.27 A: D, C, A, B

7.28 Q: A 3-kilogram café sign is hung from a 1-kilogram horizontal pole as shown in the diagram. A wire is attached to prevent the sign from rotating. Find the tension in the wire.

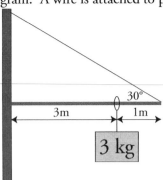

30°

3m   1m

3 kg

7.28 A: Start by drawing a diagram of the horizontal pole, showing all the forces on the pole as a means to illustrate the various torques. Assume the pivot is the attachment point on the left hand side of the pole.

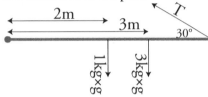

2m

3m

T

30°

1kg×g   3kg×g

Since the pole is in equilibrium, the net torque must be zero (note that torques which would cause a counter-clockwise angular acceleration are positive, and those which would contribute to a clockwise angular acceleration are negative by convention).

$$\tau_{net} = T\sin 30° \,(4m) - (3kg)\,(g)\,(3m) - (1kg)\,(g)\,(2m) = 0 \rightarrow T = \frac{(11kg \bullet m)\,(g)}{4m\sin 30°} = 54\ N$$

7.29 Q: A system of three wheels fixed to each other is free to rotate about an axis through its center. Forces are exerted on the wheels as shown. What is the magnitude of the net torque on the wheels?

7.29 A: $\tau_{net} = (-2F)\,(2R) + (2F)\,(R) + (3F)\,(1.5R) = 2.5FR$

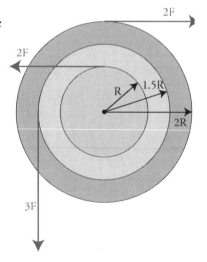

7.30 Q: A 50-kg boy and a 40-kg girl sit on opposite ends of a 3-meter see-saw. How far from the girl should the fulcrum be placed in order for the boy and girl to balance on opposite ends of the see-saw?

7.30 A: In order for the children to balance, the net torque about the fulcrum must be zero, resulting in no angular acceleration. A diagram such as the one below may be helpful in developing the mathematical relationships.

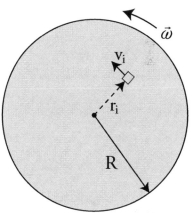

$$\vec{\tau}_{net} = 0 \rightarrow (50g)(3-x) - (40g)(x) = 0 \xrightarrow{g=10^{m}/_{s^2}} 500(3-x) - 400x = 0 \rightarrow x = 1.67\ m$$

Rotational Kinetic Energy

When kinematics was first introduced, kinetic energy was defined as the ability of a moving object to move another object. Then, translational kinetic energy for a moving object was calculated using the formula:

$$K = \tfrac{1}{2}mv^2$$

Since an object which is rotating also has the ability to move another object, it, too, must have kinetic energy. Consider the case of a uniform disc of mass m and radius R rotating about its center of mass with angular velocity ω as shown below.

You can find the kinetic energy of the entire rotating disc by treating the disc as a collection of infinitessimally small pieces of disc with mass m_i, each with velocity v_i, located a distance r_i from the axis of rotation. The kinetic energy of each little piece of the disc can be determined as $K_i = \tfrac{1}{2}m_i v_i^2 \xrightarrow{v_i = \omega r_i} K_i = \tfrac{1}{2}m_i \omega^2 r_i^2$.

You can obtain the total kinetic energy of the rotating disc by summing up all the individual kinetic energies from each of the pieces of the disc:

$$K_{rotational} = \sum_i K_i = \sum_i \tfrac{1}{2}m_i \omega^2 r_i^2 = \frac{\omega^2}{2}\sum_i m_i r_i^2 \xrightarrow{I = \sum mr^2} K_{rotational} = \tfrac{1}{2}I\omega^2$$

Rotational kinetic energy can be calculated using the analog to the translational kinetic energy formula — all you have to do is replace inertial mass with moment of inertia, and translational velocity with angular velocity!

$$K_{rot} = \tfrac{1}{2}I\omega^2$$

If an object exhibits both translational motion and rotational motion, the total kinetic energy of the object can be found by adding the translational kinetic energy and the rotational kinetic energy:

$$K_{total} = K_{translational} + K_{rotational} = \tfrac{1}{2}mv^2 + \tfrac{1}{2}I\omega^2$$

Because you're solving for energy, of course the answers will have units of Joules.

7.31 Q: Gina rolls a bowling ball of mass 7 kg and radius 10.9 cm down a lane with a velocity of 6 m/s. Find the rotational kinetic energy of the bowling ball, assuming it does not slip.

7.31 A: To find the rotational kinetic energy of the bowling ball, you need to know its moment of inertia and its angular velocity. Assume the bowling ball is a solid sphere to find its moment of inertia.
$$I = \tfrac{2}{5}mR^2 = \tfrac{2}{5}(7kg)(0.109m)^2 = 0.0333 \; kg \cdot m^2$$

Next, find the ball's angular velocity.
$$\omega = \frac{v}{r} = \frac{6 \, m/s}{0.109m} = 55 \; rad/s$$

Finally, solve for the rotational kinetic energy of the bowling ball.
$$K_{rot} = \tfrac{1}{2}I\omega^2 = \tfrac{1}{2}(0.0333 kg \cdot m^2)(55 \, rad/s)^2 = 50.4 \; J$$

7.32 Q: Find the total kinetic energy of the bowling ball from the previous problem.

7.32 A: The total kinetic energy is the sum of the translational kinetic energy and the rotational kinetic energy of the bowling ball.
$$K_{total} = \tfrac{1}{2}mv^2 + \tfrac{1}{2}I\omega^2 = \tfrac{1}{2}(7kg)(6 \, m/s)^2 + 50.4 \; J = 176 \; J$$

7.33 Q: Harrison kicks a soccer ball which rolls across a field with a velocity of 5 m/s. What is the ball's total kinetic energy? You may assume the ball doesn't slip, and treat it as a hollow sphere of mass 0.43 kg and radius 0.11 meter.

7.33 A: Immediately note that the ball will have both translational and rotational kinetic energy. Therefore, you'll need to know the ball's mass (given), translational velocity (given), moment of inertia (unknown), and rotational velocity (unknown).

Start by finding the moment of inertia of the ball, modeled as a hollow sphere.
$$I = \tfrac{2}{3}mR^2 = \tfrac{2}{3}(0.43kg)(0.11m)^2 = 0.00347 \; kg \cdot m^2$$

Next, find the rotational velocity of the soccer ball.
$$\omega = \frac{v}{r} = \frac{5 \, m/s}{0.11m} = 45.5 \; rad/s$$

Now you have enough information to calculate the total kinetic energy of the soccer ball.
$$K = \tfrac{1}{2}mv^2 + \tfrac{1}{2}I\omega^2 = \tfrac{1}{2}(0.43kg)(5 \, m/s)^2 + \tfrac{1}{2}(0.00347 kg \cdot m^2)(45.5 \, rad/s)^2 \rightarrow$$
$$K = 8.96 \; J$$

7.34 Q: Find the speed of a disc of radius R which starts at rest and rolls without slipping down an incline of height H.

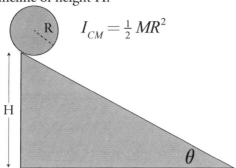

$$I_{CM} = \tfrac{1}{2} MR^2$$

7.34 A: Use a conservation of energy approach, recognizing that the gravitational potential energy of the disc at the top of incline must equal the kinetic energy of the disc at the bottom of the incline.

$$K_i + U_i = K_f + U_f \xrightarrow[U_f = 0]{K_i = 0} U_i = K_f$$

Next, recognize that the kinetic energy has translational and rotational components, then solve for the final velocity of the center of mass, utilizing the relationship between angular velocity and translational velocity.

$$U_i = K_f \rightarrow MgH = \tfrac{1}{2} Mv_{cm}^2 + \tfrac{1}{2} I\omega^2 \xrightarrow{I = \tfrac{1}{2} MR^2} MgH = \tfrac{1}{2} Mv_{cm}^2 + \tfrac{1}{2} \left(\tfrac{1}{2} MR^2 \right) \omega^2 \rightarrow$$

$$gH = \tfrac{1}{2} v_{cm}^2 + \tfrac{1}{2} \left(\tfrac{1}{2} R^2 \right) \omega^2 \xrightarrow{v = \omega R} gH = \tfrac{1}{2} v_{cm}^2 + \tfrac{1}{4} v_{cm}^2 \rightarrow gH = \tfrac{3}{4} v_{cm}^2 \rightarrow v_{cm} = \sqrt{\tfrac{4}{3} gH}$$

7.35 Q: A ball and a block of equal mass are situated on identical ramps. The objects are released from the same height. The ball rolls without slipping, and the block travels without friction. After leaving the ramp, which object travels higher and why?

A) The ball travels higher because it leaves the ramp with a higher speed.
B) The ball travels higher because it gains rotational kinetic energy alongs its path.
C) The block travels higher because it experiences no rotation.
D) The block travels higher because energy is not conserved in a rolling system.

7.35 A: (C) The block travels higher because it experiences no rotation. The ball spins as it rolls down the ramp, converting its gravitational potential energy to rotational kinetic energy and translational kinetic energy. The block, on the other hand, doesn't spin; therefore, all of its gravitational potential energy is converted into translational kinetic energy. The block therefore leaves the ramp with more speed at the same angle as the ball and therefore travels higher.

7.36 Q: A block of mass M rests on a frictionless horizontal surface attached to a light string. The string is attached to a hanging block of mass m by a pulley of mass m_p and radius R.

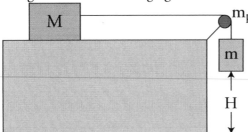

After the hanging block has fallen a distance H, what is speed of the large block?

7.36 A: Model the pulley as a uniform solid disk, so that its moment of inertia is $I = \frac{1}{2}m_p R^2$. You can then use conservation of energy to solve for the speed of the blocks (recognizing the speed of the hanging block will be the same as the speed of the larger block).

$$U_{g_i} = K_{trans_M} + K_{trans_m} + K_{rot_p} \rightarrow mgH = \tfrac{1}{2}Mv^2 + \tfrac{1}{2}mv^2 + \tfrac{1}{2}I\omega^2 \xrightarrow[\omega=\frac{v}{R}]{I=\frac{1}{2}m_p R^2}$$

$$mgH = \tfrac{1}{2}(M+m)v^2 + \tfrac{1}{2}\left(\tfrac{1}{2}m_p R^2\right)\left(\frac{v^2}{R^2}\right) \rightarrow 2mgH = (M+m)v^2 + \tfrac{1}{2}m_p v^2 \rightarrow$$

$$2mgH = v^2\left(M+m+\frac{m_p}{2}\right) \rightarrow v^2 = \frac{2mgH}{M+m+\dfrac{m_p}{2}} \rightarrow v = \sqrt{\frac{4mgH}{2M+2m+m_p}}$$

7.37 Q: A hula hoop of mass M and radius R is nailed in place so it can pivot freely as shown in the diagram below. The hoop is released and allowed to pivot freely around the nail. What is the maximum translational speed of point X on the hoop?

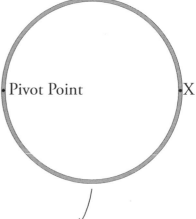

7.37 A: First find the moment of inertia of the hoop about a point on its edge using the parallel axis theorem.

$$I_p = I_{cm} + MR^2 \xrightarrow{I_{cm}=MR^2} I_p = 2MR^2$$

Next, you can apply conservation of energy to find the angular speed of point X at the bottom of its arc, and then the translational speed of point X by recognizing that the change in height of the center of mass of the hoop is the radius, R.

$$U_{g_i} = K_f \rightarrow MgR = \tfrac{1}{2}I\omega^2 \xrightarrow[\omega=\frac{v}{R}]{I=2MR^2} MgR = \tfrac{1}{2}(2MR^2)\left(\frac{v}{R}\right)^2 \rightarrow MgR = (MR^2)\left(\frac{v^2}{R^2}\right) \rightarrow$$

$$gR = v^2 \rightarrow v = \sqrt{gR}$$

Rotational Dynamics

In the chapter on dynamics, you learned about forces causing objects to accelerate. The larger the net force, the greater the linear (or translational) acceleration, and the larger the mass of the object, the smaller the translational acceleration.

$$\vec{F}_{net} = m\vec{a}$$

The rotational equivalent of this law, **Newton's 2nd Law for Rotation**, relates the torque on an object to its resulting angular acceleration. The larger the net torque, the greater the rotational acceleration, and the larger the rotational inertia, the smaller the rotational acceleration:

$$\vec{\tau}_{net} = I\vec{\alpha}$$

It's important to observe just how similar these equations are. Newton's 2nd Law for Translation used net force, the rotational version uses net torque. The translational version uses translational inertia (measured by mass), while the rotational version uses rotational inertia, or moment of inertia. The translational version uses translational acceleration, while the rotational version uses rotational acceleration. Application of the law is also similar to the translational law. Let's go through lots of examples.

7.38 Q: What is the angular acceleration experienced by a uniform solid disc of mass 2 kg and radius 0.1 m when a net torque of 10 N·m is applied? Assume the disc spins about its center.

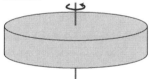

7.38 A: $\tau_{net} = I\alpha = \frac{1}{2}mR^2\alpha \rightarrow \alpha = \frac{2\tau_{net}}{mR^2} = \frac{2(10N \cdot m)}{(2kg)(0.1m)^2} = 1000\,{}^{rad}\!/\!_s$

7.39 Q: A Round-A-Bout on a playground with a moment of inertia of 100 kg·m² starts at rest and is accelerated by a force of 150N at a radius of 1m from its center. If this force is applied at an angle of 90° from the line of action for a time of 0.5 seconds, what is the final rotational velocity of the Round-A-Bout?

7.39 A: Start by making a rotational kinematics table:

Variable	Value
ω_0	0 rad/s
ω	FIND
$\Delta\theta$	?
α	?
t	0.5s

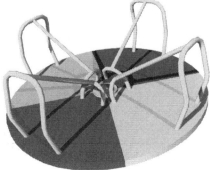

Since you only know two items on the table, you must find a third before you solve this with the rotational kinematic equations. Since you are given the moment of inertia of the Round-A-Bout as well as the applied force, you can solve for the angular acceleration using Newton's 2nd Law for Rotational Motion.

$$\tau_{net} = I\alpha \rightarrow \alpha = \frac{\tau_{net}}{I} = \frac{Fr\sin\theta}{I} = \frac{(150N)(1m)\sin 90°}{100kg \cdot m^2} = 1.5\ ^{rad}/_{s^2}$$

Now, use rotational kinematics to solve for the final angular velocity of the Round-A-Bout.

$$\omega = \omega_0 + \alpha t = 0 + (1.5\ ^{rad}/_{s^2})(0.5s) = 0.75\ ^{rad}/_{s}$$

7.40 Q: A light string attached to a mass m is wrapped around a pulley of mass m_p and radius R. If the moment of the inertia of the pulley is $\frac{1}{2}m_p R^2$, find the acceleration of the mass.

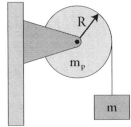

7.40 A: First draw Free Body Diagrams for both the pulley as well as the hanging mass.

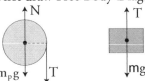

Write a Newton's 2nd Law for Rotation equation for the pulley and solve for the tension in the string.

$$\tau_{net} = I\alpha \xrightarrow{\tau = rF\sin\theta}_{I=\frac{1}{2}m_p R^2} RT = \frac{1}{2}m_p R^2 \alpha \rightarrow T = \frac{1}{2}m_p R\alpha \xrightarrow{a=R\alpha} T = \frac{1}{2}m_p a$$

Next, write a Newton's 2nd Law Equation in the y-direction for the hanging mass. Note that you can substitute in your value for T derived from the pulley FBD to solve for the acceleration.

$$F_{net_y} = ma_y \rightarrow mg - T = ma \xrightarrow{T=\frac{1}{2}m_p a} mg - \frac{1}{2}m_p a = ma \rightarrow mg = a\left(m + \frac{m_p}{2}\right) \rightarrow$$

$$a = \frac{mg}{m + \frac{m_p}{2}}$$

7.41 Q: Two blocks are connected by a light string over a pulley of mass m_p and radius R as shown in the diagram.

Find the acceleration of mass m_2 if m_1 sits on a frictionless surface.

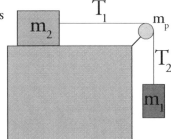

Chapter 7: Rotation

7.41 A: First draw Free Body Diagrams for the masses and the pulley.

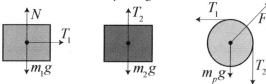

Start by writing Newton's 2nd Law equations for the x-direction for m_1, the y-direction for m_2, and rotation for the pulley. Note that T is not constant due to the pulley having inertia.

$$F_{net_x} = ma_x \rightarrow T_1 = m_1 a$$

$$F_{net_y} = ma_y \rightarrow m_2 g - T_2 = m_2 a \rightarrow T_2 = m_2 g - m_2 a$$

$$\tau_{net} = I\alpha \rightarrow T_2 R - T_1 R = I\alpha$$

Next combine the equations to eliminate the tensions, then solve for the acceleration using the relationship between angular acceleration and translational acceleration ($a = \alpha R$). Assume the pulley is a uniform disc.

$$\tau_{net} = T_2 R - T_1 R = I\alpha \xrightarrow[T_2 = m_2 g - m_2 a]{T_1 = m_1 a} R(m_2 g - m_2 a - m_1 a) = I\alpha \xrightarrow[a = \alpha R]{I_{disc} = \frac{1}{2} m_p R^2}$$

$$R(m_2 g - m_2 a - m_1 a) = \frac{1}{2} m_p R^2 \left(\frac{a}{R}\right) = \frac{1}{2} m_p aR \rightarrow m_2 g - m_2 a - m_1 a = \frac{1}{2} m_p a \rightarrow$$

$$m_2 g = m_1 a + m_2 a + \frac{1}{2} m_p a \rightarrow m_2 g = a\left(m_1 + m_2 + \frac{1}{2} m_p\right) \rightarrow a = \frac{m_2 g}{m_1 + m_2 + \frac{1}{2} m_p}$$

7.42 Q: A given force is applied to a wrench to turn a bolt of specific rotational inertia I which rotates freely about its center as shown in the following diagrams.

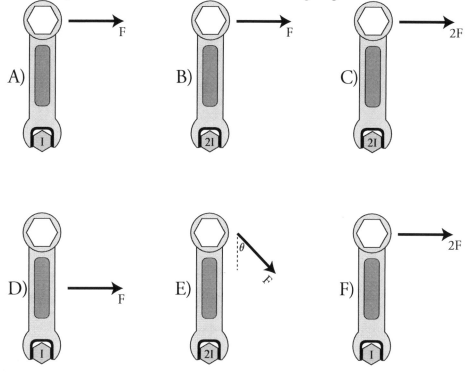

Rank the resulting angular acceleration of the bolt from greatest to least.

7.42 A: F>A=C>B=D>E

Net torque is equal to the product of the angular acceleration and the rotational inertia; therefore, angular acceleration is net torque divided by moment of inertia. Assume the length of the entire wrench is L.

Diagram	Torque	Rotational I	Angular Acceleration
A	FL	I	FL/I
B	FL	2I	FL/2I
C	2FL	2I	FL/I
D	FL/2	I	FL/2I
E	FLsin45°	2I	FL√2/4I
F	2FL	I	2FL/I

7.43 Q: A beam of mass M and length L has a moment of inertia about its center of $ML^2/12$. The beam is attached to a frictionless hinge at an angle of 45° and allowed to swing freely. Find the beam's angular acceleration.

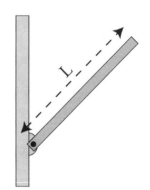

7.43 A: Begin by finding the moment of inertia of the beam about its end using the Parallel Axis Theorem.

$$I_{end} = I_{cm} + md^2 = \frac{ML^2}{12} + M\left(\frac{L}{2}\right)^2 = \frac{ML^2}{3}$$

Next, apply Newton's 2nd Law for Rotation to find the angular acceleration of the beam.

$$\tau_{net} = I\alpha \rightarrow -Mg\cos\theta\left(\frac{L}{2}\right) = I\alpha \rightarrow \alpha = \frac{-Mg\cos\theta L}{2I} \xrightarrow{I=\frac{ML^2}{3}} \alpha = \frac{-3g\cos\theta}{2L}$$

7.44 Q: A 20-kg ladder of length 8 meters sits against a frictionless wall at an angle of 60 degrees. The ladder just barely keeps from slipping.

(A) On the diagram at right, draw and label the forces acting on the ladder.
(B) Determine the force of friction of the floor on the ladder.
(C) Determine the coefficient of friction between the ladder and the floor.

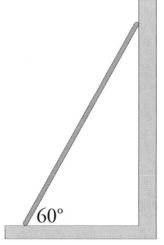

7.44 A: A)

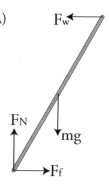

B) Utilize Newton's 2nd Law in the x- and y-directions, as well as Newton's 2nd Law for Rotation, to determine the force of Friction F_f.

$$F_{net_x} = ma_x \xrightarrow{a_x=0} F_f - F_w = 0 \rightarrow F_f = F_w$$

$$F_{net_y} = ma_y \xrightarrow{a_y=0} F_N - mg = 0 \rightarrow F_N = mg = (20kg)(10^m/_{s^2}) = 200\ N$$

$$\tau_{net} = I\alpha \xrightarrow{\alpha=0} 0 = -4mg\cos 60° + 8F_w\sin 60° \rightarrow F_w = F_f = \frac{4mg\cos 60°}{8\sin 60°} = 57.7\ N$$

C) Solve for the coefficient of friction.

$$F_f = \mu F_N \rightarrow \mu = \frac{F_f}{F_N} = \frac{57.7N}{200N} = 0.29$$

Rolling

Problems involving circular objects rolling across a surface provide a great opportunity to explore both rotational and translational motion simultaneously. When an object is rolling without slipping, you have rotational motion of the object about its axis of rotation coupled with translational motion of the center of mass through space. The key to solving these types of problems is recognizing the following key conditions hold true for objects that roll without slipping:

$$s = r\theta \quad v = r\omega \quad a = r\alpha$$

Further, the total kinetic energy of the rolling object is obtained by adding together the object's rotational kinetic energy and translational kinetic energy. Let's take a look at a couple examples.

7.45 Q: A disc of radius R rolls down an incline of angle θ without slipping.

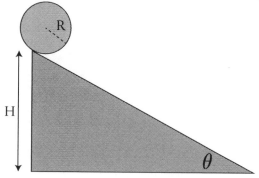

Find the force of friction on the disc.

7.45 A: Start with a free body diagram for the disc on the ramp, labeling all forces at their point of action (below left). Recognizing the weight of the disc doesn't line up with the axes, you can break the weight up into components parallel and perpendicular to the axes (below right).

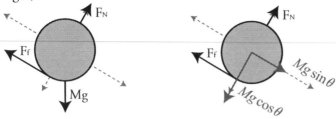

Using the pseudo-free body diagram on the right, you can write Newton's Second Law equations for the x- and y-directions.

$$F_{net_x} = Ma_x \rightarrow Mg\sin\theta - F_f = Ma_x$$

$$F_{net_y} = Ma_y \xrightarrow{y=0} F_N - Mg\cos\theta = 0 \rightarrow F_N = Mg\cos\theta$$

Next, you can apply Newton's 2nd Law for Rotation to the problem to solve for the force of friction. Note that we'll focus on just the magnitudes of the torque and angular acceleration, as they are in the same direction.

$$\tau_{net} = I\alpha \xrightarrow{I=\frac{1}{2}MR^2} F_fR = \frac{MR^2\alpha}{2} \xrightarrow{a=\alpha R} F_fR = \frac{MRa}{2} \rightarrow F_f = \frac{Ma}{2} \xrightarrow{Ma=Mg\sin\theta - F_f}$$

$$F_f = \frac{Mg\sin\theta - F_f}{2} \rightarrow 2F_f = Mg\sin\theta - F_f \rightarrow 3F_f = Mg\sin\theta \rightarrow F_f = \frac{Mg\sin\theta}{3}$$

7.46 Q: A round object of mass m and radius r sits at the top of a ramp of length L inclined at an angle of θ with the horizontal. When released, it rolls down the ramp without slipping.

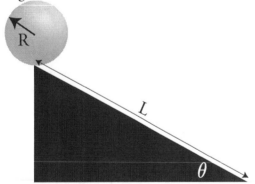

A) Derive an expression for the gravitational potential energy of the object in terms of m, L, θ and fundamental constants.

A student wishes to experimentally determine the object's moment of inertia, I, by adjusting the angle of the ramp and observing the behavior of the object as it rolls down the ramp without slipping.

B) Describe an experimental procedure the students could use to collect the necessary data, including any equipment the student would need. Your description should include a list of the independent and dependent variables.

C) Derive an expression for the velocity of the object at the bottom of the ramp in terms of the length of the ramp, L, and the time it takes to travel down the ramp, t.

D) Using the law of conservation of energy, derive an expression for the moment of inertia of the object in terms of m, r, θ, t, L, and any fundamental constants.

E) The student plots the square of the time it takes for the object to travel down the ramp, t^2, as a function of $1/\sin(\theta)$ to obtain a linear graph. How should the student determine the moment of inertia of the object from this graph? Highlight the calculation(s) used.

F) Now suppose that you were not given the radius of the object. Describe an experimental procedure you could use to determine the radius, including any equipment you would need.

G) Derive an expression for the moment of inertia of the object in terms of m, r, θ, t, L, and any fundamental constants using an alternate approach to your method from part (D).

7.46 A: A) Setting the bottom of the ramp as the zero point of gravitational potential energy:
$$U_g = mgH \xrightarrow{H = L\sin\theta} U_g = mgL\sin\theta$$

B) Using a stopwatch, the student could start the stopwatch when the object is released from rest, and halt the stopwatch when the object reaches the bottom of the incline, repeating this for varying levels of inclination and recording the independent variable (θ) and the dependent variable (t) in a data table. Other equivalent answers are acceptable, such as utilizing photogates to determine the time, or determining the speed of the object as it reaches the bottom of the incline.

C) Using kinematic equations, $v = \dfrac{2L}{t}$

D) Begin by recognizing that the gravitational potential energy at the top of the ramp must equal the total kinetic energy of the rotating object at the bottom of the ramp.
$$U_{g_{top}} = K_{bottom} \rightarrow mgL\sin\theta = \tfrac{1}{2}mv^2 + \tfrac{1}{2}I\omega^2 \rightarrow I = \frac{2mgL\sin\theta}{\omega^2} - \frac{mv^2}{\omega^2} \xrightarrow[\omega = v/r]{v = \omega r}$$
$$I = \frac{2mgL\sin\theta r^2}{v^2} - mr^2 = mr^2\left(\frac{2gL\sin\theta}{v^2} - 1\right) \xrightarrow[v^2 = 4L^2/t^2]{v = 2L/t} I = mr^2\left(\frac{2gLt^2\sin\theta}{4L^2} - 1\right) \rightarrow$$
$$I = mr^2\left(\frac{gt^2\sin\theta}{2L} - 1\right)$$

E) Plotting the function $t^2 = \left[\dfrac{2L(I + mr^2)}{mr^2 g}\right]\left(\dfrac{1}{\sin\theta}\right)$ matches the form y=mx, where the

slope of the line is given by the terms in brackets. Therefore, to find the moment of inertia, find the slope of the best-fit line from the graph. You can then rearrange the equation above to obtain the moment of inertia as $I = mr^2\left(\dfrac{g \times slope}{2L} - 1\right)$.

F) One possible answer involves wrapping a string around the circumference of the object, marking off the circumference on the string, then measuring the length of the marked area on the string when it is pulled taut into a line. The radius, then, is the circumference divided by 2π.

G) The moment of inertia may be determined by a conservation of energy approach, as shown in part (D), but you can also calculate it using a Newton's 2nd Law approach. First, write Newton's 2nd Law for Rotation, focusing on the magnitude of the quantities since both torque and angular acceleration are in the same direction.

$$\tau_{net} = I\alpha \rightarrow F_f r = I\alpha \rightarrow F_f = \frac{I\alpha}{r}$$

After creating a free body diagram and pseudo free body diagram (see previous problem), you can write Newton's 2nd Law in the x-direction, and substitute in your previous result for the force of friction.

$$F_{net_x} = ma_x \rightarrow mg\sin\theta - F_f = ma \rightarrow mg\sin\theta - \frac{I\alpha}{r} = ma \xrightarrow{a = a/r}$$

$$mg\sin\theta - \frac{I a}{r^2} = ma \rightarrow \frac{I a}{r^2} = mg\sin\theta - ma \rightarrow I = mr^2\left(\frac{g\sin\theta}{a} - 1\right) \xrightarrow{a = 2L/t^2}$$

$$I = mr^2\left(\frac{gt^2\sin\theta}{2L} - 1\right)$$

When an object skids across a surface, it may roll with slipping. In this case, the condition for rolling without slipping, $v = r\omega$, may not be applicable. These sorts of cases can be easily visualized at a bowling alley.

Imagine a bowler slings a bowling ball down an alley with backspin. The ball is rotating, and its center of mass is moving down the alley, but the $v = r\omega$ condition isn't being met. Given time, however, the friction on the ball will act on the ball in such a way that the ball will roll without slipping, meeting the $v = r\omega$ condition (assuming the alley lane is long enough).

Alternately, imagine a bowling ball is slung down the alley in such a manner that it has a tremendous forward spin, but is again skidding down the lane. In this case as well, eventually the frictional forces on the ball will adjust the angular velocity and translational velocity such that the ball rolls without slipping.

Let's take a look at an example in which a bowling ball is slung down an alley with no initial rotational motion, so that it skids some distance before rolling without slipping.

7.47* Q: A bowling ball of mass M and radius R skids horizontally without rotating down the alley with an initial velocity of v_0. Find the distance the ball skids before rolling, given a coefficient of kinetic friction between the ball and the alley surface of μ_k.

7.47* A: Begin with a free body diagram of the ball traveling down the alley, then write Newton's 2nd Law equations as shown.

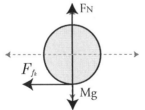

$$F_{net_y} = ma_y \rightarrow F_N - Mg = 0 \rightarrow F_N = Mg$$

$$F_{net_x} = ma_x \rightarrow -F_f = Ma \xrightarrow{F_f = \mu_k F_N} -\mu_k F_N = Ma \xrightarrow{F_N = Mg} -\mu_k Mg = Ma \rightarrow a = -\mu_k g$$

Next, you can solve for the velocity as a function of time using a kinematic equation, recognizing the linear acceleration of the ball is constant.

$$v = v_0 + at \xrightarrow{a = -\mu_k g} v = v_0 - \mu_k gt$$

With this in mind, you can then write Newton's 2nd Law for Rotation and incorporate your previous findings. Because the angular acceleration and net torque are in the same direction, you can focus on just the magnitude of the quantities.

$$\tau_{net} = I\alpha \rightarrow F_f R = I\alpha \xrightarrow[I = \frac{2}{5}MR^2]{F_f = \mu_k Mg} \mu_k MgR = \frac{2}{5}MR^2\alpha \rightarrow \mu_k g = \frac{2}{5}R\alpha \rightarrow \alpha = \frac{5\mu_k g}{2R}$$

Solving for the angular velocity using rotational kinematics, again recognizing that the angular acceleration is constant provides you:

$$\omega = \omega_0 + \alpha t \xrightarrow[a = \frac{5\mu_k g}{2R}]{\omega_0 = 0} \omega = \frac{5\mu_k g}{2R}t$$

Then, apply the condition at which the ball stops slipping to solve for the time at which this condition is met.

$$v = R\omega \xrightarrow[\omega = \frac{5\mu_k g}{2R}t]{v = v_0 - \mu_k gt} v_0 - \mu_k gt = \frac{5\mu_k g}{2}t \rightarrow v_0 = \mu_k gt + \frac{5\mu_k g}{2}t \rightarrow$$

$$v_0 = \frac{7}{2}\mu_k gt \rightarrow t = \frac{2v_0}{7\mu_k g}$$

Finally, you can use another kinematic equation to solve for the displacement of the ball at the time the ball stops slipping.

$$\Delta x = x_0 + v_0 t + \frac{1}{2}at^2 \xrightarrow[a = -\mu_k g]{t = \frac{2v_0}{7\mu_k g}} \Delta x = v_0\left(\frac{2v_0}{7\mu_k g}\right) + \frac{1}{2}(-\mu_k g)\left(\frac{2v_0}{7\mu_k g}\right)^2 \rightarrow$$

$$\Delta x = \frac{2}{7}\left(\frac{v_0^2}{\mu_k g}\right) - \frac{2}{49}\left(\frac{v_0^2}{\mu_k g}\right) = \frac{12}{49}\left(\frac{v_0^2}{\mu_k g}\right)$$

Angular Momentum

Much like linear momentum describes how difficult it is to stop an object moving linearly, **angular momentum** ($\vec{L}$) is a vector describing how difficult it is to stop a rotating object. The total angular momentum of a system is the sum of the individual angular momenta of the objects comprising that system.

A mass with velocity $\vec{v}$ moving at some position $\vec{r}$ about point Q has angular momentum $\vec{L}_Q$, as shown in the diagram at right. Note that angular momentum depends on the point of reference.

$$\vec{L}_Q = \vec{r} \times \vec{p} = \vec{r} \times m\vec{v} = (\vec{r} \times \vec{v})m$$

The magnitude of the angular momentum about point Q can be found from the definition of the cross product, and the direction given by a right-hand rule. Point the fingers of your right hand in the direction of the position vector, and bend your fingers in the direction of the momentum (or velocity) vector. Your thumb will then point in the direction of the angular momentum vector, which is perpendicular to both the position vector and the momentum vector.

$$|\vec{L}_Q| = mvr\sin\theta \xrightarrow{v = \omega r} |\vec{L}_Q| = mr^2\omega$$

Note how the formula for the magnitude of the angular momentum about point Q contains the term mr², the same term that would describe the moment of inertia of a point mass moving around point Q. For an object rotating about its center of mass, the angular momentum of the object is equal to the product of the object's moment of inertia and its angular velocity.

$$\vec{L} = I\vec{\omega}$$

This is an intrinsic property of an object rotating about its center of mass, and is known as the object's **spin angular momentum**. It is constant regardless of your reference point.

7.48 Q: Find the angular momentum of a planet of mass m in a perfectly circular orbit about a sun with angular velocity ω.

7.48 A: Taking the angular momentum about point Q, located at the center of the circular orbit:

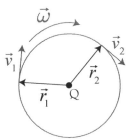

$$\vec{L}_Q = \vec{r} \times \vec{p} \rightarrow |\vec{L}_Q| = mvr\sin\theta \xrightarrow{\theta=90°} |\vec{L}_Q| = mvr$$

Utilizing the right-hand rule, you can find the direction of the angular momentum vector as into the plane of the page.

Alternately, you could also find the angular momentum by treating the orbiting planet as a point particle with a moment of inertia of mr² with respect to point Q. The direction of the angular momentum vector is the same as the rotational velocity vector (into the page), and its magnitude is gain given by mvr.

$$\vec{L}_Q = I\vec{\omega} \xrightarrow{I=mr^2} \vec{L}_Q = mr^2\vec{\omega} \xrightarrow{v=\omega r} |\vec{L}_Q| = mvr$$

7.49 Q: Find the magnitude of the angular momentum for a 5 kg point particle located at (2,2) with a velocity of 2 m/s east
A) about the origin (point O)
B) about point P at (2,0)
C) about point Q at (0,2)

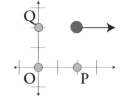

7.49 A: A) $|\vec{L}_O| = mvr\sin\theta = (5kg)(2\,{}^m/_s)(2\sqrt{2}\,m)\left(\frac{\sqrt{2}}{2}\right) = 20\,{}^{kg\cdot m^2}/_s$

B) $|\vec{L}_P| = mvr\sin\theta = (5kg)(2\,{}^m/_s)(2m)(1) = 20\,{}^{kg\cdot m^2}/_s$

C) $|\vec{L}_Q| = mvr\sin\theta = (5kg)(2\,{}^m/_s)(2m)(0) = 0$

7.50 Q: Four particles, each of mass M, move in the x-y plane with varying velocities as shown in the diagram. The velocity vectors are drawn to scale. Rank the magnitude of the angular momentum about the origin for each particle from largest to smallest.

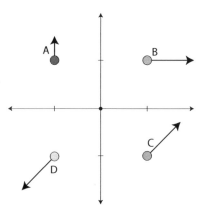

7.50 A: C, B, A, D. C has the highest angular momentum as it has the maximum velocity and the angle between the position vector to the mass and the velocity vector is 90 degrees. B has the next highest angular momentum as it has the maximum velocity, and the angle between the position vector to the mass and the velocity vector is 45 degrees. A has the third highest angular momentum as it has half the maximum velocity, and the angle between the position vector to the mass and the velocity vector is 45 degrees. D has the lowest angular momentum (0) about the origin since the position vector to the mass and its velocity are parallel.

7.51 Q: A disc rotates clockwise about its axis as shown in the diagram. The direction of the angular momentum vector is:
A) out of the plane of the page
B) into the plane of the page
C) toward the top of the page
D) toward the bottom of the page

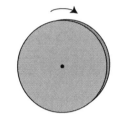

7.51 A: (B) into the plane of the page. The direction of the angular momentum vector is given by the right-hand rule. In this case, wrap the fingers of your right-hand around the axle in the direction of the rotational velocity, and your thumb points in the direction of the angular momentum vector.

In thinking about angular momentum, consider a situation in which an object spins about its axis. In the absence of any external torques, you would expect the angular momentum to remain constant. However, a torque on the object could change the object's angular momentum. Let's develop the mathematical relationship between torque and angular momentum by starting with the definitions of angular momentum and cross products.

$$\vec{L}_Q = \vec{r} \times \vec{p} \xrightarrow{\frac{d}{dt}(\vec{A} \times \vec{B}) = \frac{d\vec{A}}{dt} \times \vec{B} + \vec{A} \times \frac{d\vec{B}}{dt}} \frac{d\vec{L}_Q}{dt} = \frac{d\vec{r}}{dt} \times \vec{p} + \vec{r} \times \frac{d\vec{p}}{dt} \xrightarrow[\frac{d\vec{p}}{dt} = \vec{F}]{\frac{d\vec{r}}{dt} = \vec{v}}$$

$$\frac{d\vec{L}_Q}{dt} = \vec{v} \times \vec{p} + \vec{r} \times \vec{F} \xrightarrow{\vec{v} \times \vec{p} = 0 \ (same\ direction)} \frac{d\vec{L}_Q}{dt} = \vec{r} \times \vec{F} \xrightarrow{\vec{r} \times \vec{F} = \vec{\tau}} \frac{d\vec{L}_Q}{dt} = \vec{\tau}_Q$$

This verifies that a torque on an object changes the object's angular momentum, and a change in angular momentum is caused by a torque.

Conservation of Angular Momentum

Previously, you learned that linear momentum, the product of an object's mass and its velocity, is conserved in a closed system. In similar fashion, **spin angular momentum**, the product of an object's moment of inertia and its angular velocity about the center of mass, is also conserved in a closed system with no external net torques applied. This is the law of **conservation of angular momentum**.

This can be observed by watching a spinning figure skater. As the skater launches into a spin, she generates rotational velocity by applying a torque to her body. The skater now has an angular momentum as she spins around an axis which is equal to the product of her moment of inertia (rotational inertia) and her rotational velocity ($L = I\omega$).

To increase the rotational velocity of her spin, she pulls her arms in close to her body, reducing her moment of inertia. Angular momentum is conserved, therefore rotational velocity must increase ($L = I\boldsymbol{\omega}$). Then, before coming out of the spin, the skater reduces her rotational velocity by move her arms away from her body, increasing her moment of inertia.

Let's examine a number of example problems involving angular momentum.

7.52 Q: An ice skater spins with a specific angular velocity. She brings her arms and legs closer to her body, reducing her moment of inertia to half its original value. Describe what happens to her:
A) angular momentum
B) angular velocity
C) rotational kinetic energy

7.52 A: A) angular momentum remains constant
B) angular velocity is doubled
C) rotational kinetic energy is doubled ($K = \frac{1}{2}I\omega^2$)

7.53 Q: If the rotational kinetic energy of the skater doubles in the previous problem, doesn't that violate the law of conservation of energy? Explain.

7.53 A: This does not violate the law of conservation of energy. In order to pull her arms and legs inward, the skater has to do work. This work done accounts for the change in the rotational kinetic energy of the skater.

7.54 Q: A disc with moment of inertia 1 kg•m² spins about an axle through its center of mass with angular velocity 10 radians per second. An identical disc which is not rotating is slid along the axle until it makes contact with the first disc.

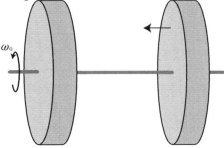

If the two discs stick together, what is their combined angular velocity?

7.54 A: Utilize conservation of angular momentum:
$$L_0 = L_f \rightarrow I_0\omega_0 = I_f\omega_f \rightarrow \omega_f = \frac{I_0\omega_0}{I_f} = \frac{(1kg \cdot m^2)(10^{rad}/_s)}{(2kg \cdot m^2)} \rightarrow \omega_f = 5^{rad}/_s$$

7.55 Q: Angelina spins on a rotating pedestal with an angular velocity of 8 radians per second. Bob throws her an exercise ball, which increases her moment of inertia from 2 kg·m² to 2.5 kg·m². What is Angelina's angular velocity after catching the exercise ball? (Neglect any external torque from the ball.)

7.55 A: Since there are no external torques, you know that the initial spin angular momentum must equal the final spin angular momentum, and can therefore solve for Angelina's final angular velocity:

$$L_0 = L_f \rightarrow I_0 \omega_0 = I_f \omega_f \rightarrow \omega_f = \frac{I_0 \omega_0}{I_f} = \frac{(2kg \cdot m^2)(8^{rad}/_s)}{(2.5kg \cdot m^2)} \rightarrow \omega_f = 6.4^{rad}/_s$$

7.56 Q: A constant force F is applied for a constant time at various points of the object below, as shown in the diagram.

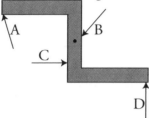

Rank the magnitude of the change in the object's angular momentum due to the force from smallest to largest.

7.56 A: B, C, A, D.

7.57 Q: Jean stands at the exact center of a large spinning frictionless uniform disk of mass M and radius R with moment of inertia $I = \frac{1}{2}MR^2$. As she walks from the center to the edge of the disk, the angular speed of the disk is quartered. Which of the following statements is true?
A) Jean's mass is less than the mass of the disk.
B) Jean's mass is equal to the mass of the disk.
C) Jean's mass is between the mass of the disk and twice the mass of the disk.
D) Jean's mass is more than twice the mass of the disk.

7.57 A: (C) Jean's mass is between the mass of the disk and twice the mass of the disk. The initial moment of inertia of the system is approximately $I = \frac{1}{2}MR^2$. Jean's mass (let's call it m) does not contribute significantly to this moment of inertia as she stands at the exact center of the large disk. As she walks from the center of the disk to the edge of the disk, however, she adds her moment of inertia to that of the disk, so that the total moment of inertia of the system is now $I = \frac{1}{2}MR^2 + mR^2$. Calling the initial angular speed of the disk ω and applying the law of conservation of angular momentum, find Jean's mass as follows:

$$L_i = L_f \rightarrow \frac{1}{2}MR^2 \omega = (\frac{1}{2}MR^2 + mR^2)\frac{\omega}{4} \rightarrow 4M = M + 2m \rightarrow m = \frac{3}{2}M$$

7.58 Q: A spinning plate in a microwave with moment of inertia I rotates about its center of mass at a constant angular speed ω. When the microwave ends its cook cycle, the plate comes to rest in time Δt due to a constant frictional force F applied a distance r from the axis of rotation. What is the magnitude of the frictional force F?

A) $F = \dfrac{I\omega}{r\Delta t}$ C) $F = \dfrac{\omega r^2}{I\Delta t}$

B) $F = \dfrac{Ir}{\omega\Delta t}$ D) $F = \dfrac{Ir^2}{\omega\Delta t}$

7.58 A: (A) The change in angular momentum of the plate is given by the product of the net torque and the time interval over which it is applied.

$$\Delta L = I\Delta\omega = \tau\Delta t \to I\omega = Fr\Delta t \to F = \dfrac{I\omega}{r\Delta t}$$

7.59 Q: Marianna races her bike across a horizontal path. Suddenly, a squirrel runs in front of her. Marianna slams on both her front and rear brakes, which results in the bike flipping over the front wheel and Marianna flying over the handle bars.

Which of the following does NOT contribute to an explanation of why the bike flips and Marianna flies over the handlebars?

A) Marianna has a tendency to continue moving at a constant velocity, so while the bike stops, Marianna continues her previous motion.
B) Conservation of angular momentum of the bike and wheels indicates that if the wheels stop spinning in one direction, the bike must spin in the opposite direction.
C) The large negative acceleration of the bike/rider system reduces the moment of inertia of the system, increasing the system's angular acceleration and causing a rotation of the bike.
D) The force of the applied brakes at a distance from the center of mass of the bike and rider produces a net torque on the bike, causing a rotation bringing the back wheel of the bike up.

7.59 A: (C) The large negative acceleration of the bike/rider system reduces the moment of inertia of the system, increasing the system's angular acceleration and causing a rotation of the bike.

(A) is true, as a restatement of Newton's 1st Law of Motion. (B) is true, though the effect may be rather small if the mass of the wheels is relatively small compared to the mass of the rest of the bike. (D) is true as the force of friction provides a net torque on the bike (similar to how a motorcycle may "pop a wheelie" when accelerated quickly). (C) is completely made up, however, and is incorrect.

7.60 Q: A hoop with moment of inertia 0.1 kg•m² spins about a frictionless axle with an angular velocity of 5 radians per second. At what radius from the center of the hoop should a force of 2 newtons be applied for 3 seconds in order to accelerate the hoop to an angular speed of 10 radians per second?
A) 8.3 cm
B) 12.5 cm
C) 16.7 cm
D) 25 cm

7.60 A: (A) 8.3 cm. A net torque will change the angular momentum of the hoop. Solving for the distance at which the force must be applied to create the appropriate torque:

$$\tau_{net} = I\alpha \rightarrow Fr = I\left(\frac{\omega_f - \omega_i}{t}\right) \rightarrow r = \frac{I}{F}\left(\frac{\omega_f - \omega_i}{t}\right) = \frac{0.1}{2}\left(\frac{10 - 5}{3}\right) = 0.083 \, m = 8.3 \, cm$$

7.61 Q: A particle of mass m is launched with velocity v toward a uniform disk of mass M and radius R which can rotate about a point on its edge as shown. The disk is initially at rest. After the particle strikes the edge of the disk and sticks, the magnitude of the final angular velocity of the disk-particle system is given by: $|\vec{\omega}| = \left(\frac{4m}{8m + 3M}\right)\left(\frac{v}{R}\right)$.

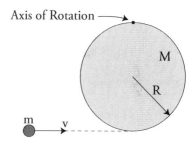

Determine the moment of inertia of the disk about its axis of rotation.

7.61 A: If you know and apply the parallel axis theorem, this is a very straightforward problem:
$$I_{total} = I_{disc_{cm}} + md^2 = \frac{1}{2}MR^2 + MR^2 = \frac{3}{2}MR^2.$$

If you don't apply the parallel axis theorem, it becomes considerably more complex, but is a great exercise in applying the law of conservation of angular momentum. Treating the disc and particle as a single system, the total angular momentum of the system must be conserved in the absence of external torques, therefore the initial angular momentum of the system must equal the final angular momentum of the system.
$$L_i = L_f = I\omega_f$$
Substituting in the initial angular momentum of the particle and the final angular velocity of the system and solving for the total momentum of inertia of the disc-particle system:

$$L_i = I\omega_f \rightarrow mv(2R) = I_{total}\omega_f \xrightarrow{\omega_f = \left(\frac{4m}{8m+3M}\right)\left(\frac{v}{R}\right)} 2mvR = I_{total}\left(\frac{4m}{8m+3M}\right)\left(\frac{v}{R}\right) \rightarrow$$

$$I_{total} = 4mR^2 + \tfrac{3}{2}MR^2$$

Finally, recognizing the total moment of inertia of the system is the sum of the moment of inertia of the particle and the moment of inertia of the disk, solve for the moment of inertia of the disk.

$$I_{disk} = I_{total} - I_{particle} = \left(4mR^2 + \tfrac{3}{2}MR^2\right) - \left(m(2R)^2\right) = \tfrac{3}{2}MR^2 \rightarrow I_{disk} = \tfrac{3}{2}MR^2$$

Translational vs. Rotational Motion

To this point, we've developed a number of analogs, or similarities, between translational and rotational motion. With just a little more work, we can extend those similarities to give a more complete picture of the relationships between translational and rotational motion.

Variable	Translational	Angular
Displacement	$\Delta \vec{r}$	$\Delta \vec{\theta}$
Velocity	$\vec{v}$	$\vec{\omega}$
Acceleration	$\vec{a}$	$\vec{\alpha}$
Time	t	t
Force/Torque	$\vec{F}$	$\vec{\tau} = \vec{r} \times \vec{F}$
Mass/Moment of Inertia	m	$I = \int r^2\, dm$

Variable	Translational	Angular
Displacement	$s = r\theta$	$\theta = \dfrac{s}{r}$
Velocity	$v = r\omega$	$\omega = \dfrac{v}{r}$
Acceleration	$a = r\alpha$	$\alpha = \dfrac{a}{r}$
Kinematics	$\vec{a} = \dfrac{d\vec{v}}{dt} = \dfrac{d^2\vec{r}}{dt^2}$	$\vec{\alpha} = \dfrac{d\vec{\omega}}{dt} = \dfrac{d^2\vec{\theta}}{dt^2}$
Force/ Torque	$\vec{F}_{net} = m\vec{a} = \dfrac{d\vec{p}}{dt}$	$\vec{\tau}_{net} = I\vec{\alpha} = \dfrac{d\vec{L}}{dt}$
Momentum/ Angular Momentum	$\vec{p} = m\vec{v}$	$\vec{L} = \vec{r} \times \vec{p} = I\vec{\omega}$
Impulse	$\vec{J} = \int \vec{F}\,dt = \Delta\vec{p}$	$\Delta\vec{L} = \int \vec{\tau}\,dt$
Kinetic Energy	$K_{trans} = \tfrac{1}{2}mv_{cm}^2$	$K_{rot} = \tfrac{1}{2}I\omega^2$
Work	$W = \int \vec{F}\cdot d\vec{l} = -\Delta U$	$W = \int \tau\,d\theta = -\Delta U$
Power	$P = \dfrac{dW}{dt} = F\dfrac{dl}{dt} = Fv$	$P = \dfrac{dW}{dt} = \tau\dfrac{d\theta}{dt} = \tau\omega$

Chapter 7: Rotation

Let's take a look at a few more problems that combine these concepts as well as application of previous fundamental principles.

7.62 Q: A wagon wheel is comprised of a hoop of mass 6 kg at a radius 0.3 meter from the center point, and is supported by 6 spokes, each of which is a uniform solid rod of mass 1 kg. The wheel accelerates from rest to an angular speed of 6 radians per second in a time of 12 seconds by application of a constant force applied at a right angle to the spokes at the point shown.
A) Determine the moment of inertia of the wheel.
B) Determine the angular acceleration of the wheel.
C) Determine the net torque on the wheel.
D) Determine the force applied.
E) Determine the average power input to the wheel.

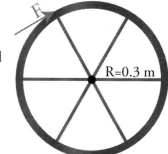

R=0.3 m

7.62 A: A) $I_{total} = I_{hoop} + I_{spokes} = MR^2 + 6\left(\frac{mR^2}{3}\right) = (6kg)(0.3m)^2 + 6\left(\frac{(1kg)(0.3m)^2}{3}\right) \rightarrow$

$I = 0.72 \; kg \cdot m^2$

B) $\alpha = \frac{\omega_f - \omega_i}{t} = \frac{6^{rad}/_s}{12s} = 0.5^{rad}/_{s^2}$

C) $\tau_{net} = \frac{\Delta L}{t} = \frac{I\omega_f - 0}{t} = \frac{(0.72kg \cdot m^2)(6^{rad}/_s)}{12s} = 0.36 \; N \cdot m$

D) $\vec{\tau} = \vec{F} \times \vec{r} \rightarrow |\vec{\tau}| = |\vec{F}||\vec{r}|\sin\theta \xrightarrow{\theta = 90°} \tau = Fr \rightarrow F = \frac{\tau}{r} = \frac{0.36N \cdot m}{0.3m} = 1.2 \; N$

E) $P = \tau\omega_{avg} = (0.36N \cdot m)(3^{rad}/_s) = 1.08 \; W$

7.63 Q: Sue spins a pottery wheel with unknown moment of inertia about its center of mass with an initial angular velocity $\vec{\omega}_i$. A ball of clay with mass m is added to the wheel at a distance d from the wheel's center of mass, slowing the wheel to a final angular velocity of $\vec{\omega}_f$. What is the moment of inertia of the pottery wheel in terms of m, d, ω_i, and ω_f? Assume the ball of clay may be modeled as a point particle and added to the wheel in such a manner as to exhibit no net torque on the wheel.

7.63 A: Assuming no net external torques are applied, the angular momentum of the system must be conserved.

$L_i = L_f \rightarrow I_i\omega_i = I_f\omega_f \xrightarrow{I_{clay} = md^2} I_i\omega_i = (I_i + md^2)\omega_f \rightarrow I_i\omega_i = I_i\omega_f + md^2\omega_f \rightarrow$

$I_i\omega_i - I_i\omega_f = md^2\omega_f \rightarrow I_i(\omega_i - \omega_f) = md^2\omega_f \rightarrow I_i = \frac{md^2\omega_f}{(\omega_i - \omega_f)}$

7.64 Q: In outer space, a ball of putty strikes a uniform solid rod as shown below.

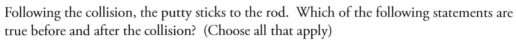

Following the collision, the putty sticks to the rod. Which of the following statements are true before and after the collision? (Choose all that apply)
A) Mass of the system is conserved
B) Translational kinetic energy of the system is conserved.
C) Rotational kinetic energy of the system is conserved.
D) Total kinetic energy of the system is conserved.
E) Total mechanical energy of the system is conserved.
F) Total energy of the system is conserved.
G) Linear momentum of the system is conserved.
H) Angular momentum of the system is conserved.
I) Net force on the system is conserved.
J) Net torque on the system is conserved.

7.64 A: A, F, G, H. Mass, total energy, linear momentum, and angular momentum are conserved. Kinetic energy is not conserved as this is an inelastic (sticky) collision, and total mechanical energy of the system is not conserved as some energy is converted to non-mechanical forms in an inelastic collision.

Chapter 8: Oscillations

Objectives

1. Describe the conditions necessary for simple harmonic motion.
2. Write an equation of the form $A\cos(\omega t)$ or $A\sin(\omega t)$ describing simple harmonic motion.
3. Recognize simple harmonic motion when expressed in differential equation form.
4. Calculate the kinetic and potential energies of an oscillating system.
5. Analyze problems involving horizontal and vertical springs attached to masses.
6. Determine the period of oscillations for systems involving combinations of springs.
7. Derive the expression for the period of both a simple and physical pendulum.
8. Understand the application and limits of the small angle approximation in deriving the period of a pendulum.
9. Identify situations in which objects will resonate in response to an external force.

"Since I was a kid, I always wanted to figure out how to make a bass line that was a pendulum - like, gravity would control it, and then you could make it play different notes."

—Bjork

Simple Harmonic Motion

W̲hen an object is displaced, it may be subject to a restoring force, resulting in a periodic oscillating motion. If the displacement is directly proportional to the linear restoring force, the object undergoes **simple harmonic motion** (SHM). Examples of phenomena which can be modeled by simple harmonic motion include a mass on a spring, the pendulum on a grandfather clock, a tree limb oscillating after you brush it walking through the woods, a child on a swing, the strings vibrating on a guitar, the electrical current powering your computer, even the vibrations of atoms in a solid can be modeled as simple harmonic motion.

Uniform circular motion also has ties to simple harmonic motion. Consider an object moving in a horizontal circle of radius A at constant angular velocity $\vec{\omega}$ as shown in the diagram below. At any given point in time, the x-position of the object can be described by $x = A\cos\theta$, and the y-position can be described by $y = A\sin\theta$.

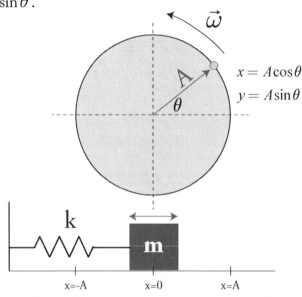

Now imagine a block on a frictionless surface attached to a wall by a spring. If that block is displaced from its equilibrium position by an amount A, and released, its simple harmonic motion along the frictionless surface will mirror the x-motion of the object moving in uniform circular motion. Simple harmonic motion and uniform circular motion are very closely related, as simple harmonic motion describes one dimension of an object moving in uniform circular motion!

We can use this analogy to further develop equations describing simple harmonic motion. Let's start with the relationship between angular velocity and angular position, recalling that angular velocity is the time rate of change of angular position (the derivative of angular position with respect to time). By separation of variables and a bit of integration, we can determine angular position as a function of time, which will come in handy very shortly.

$$\omega = \frac{d\theta}{dt} \rightarrow \int_{t=0}^{t} \omega dt = \int_{\theta=0}^{\theta} d\theta \rightarrow \omega t = \theta$$

Next, let's take a look at the mass attached to the spring, depicted above. Applying Newton's 2nd Law in the x-direction and assuming a linear spring that obeys Hooke's Law, we can derive a general relationship for simple harmonic motion.

$$F_{net_x} = ma_x = -kx \xrightarrow{a = \frac{d^2x}{dt^2}} m\frac{d^2x}{dt^2} = -kx \rightarrow \frac{d^2x}{dt^2} + \frac{k}{m}x = 0$$

Note that this differential equation has the second derivative of a function x, added to that function itself, to provide a final value of zero. The solution to this problem, therefore, must involve a function that, when added to its second derivative, gives a value of zero. Both the sine and cosine functions meet this requirement. A possible solution, then is:

$$x(t) = A\cos\left(\sqrt{\frac{k}{m}}\,t\right)$$

In this solution, the radical term, $\sqrt{\frac{k}{m}}$, represents the **angular frequency** of the motion, which is represented with the symbol ω.

$$\omega = \sqrt{\frac{k}{m}} \rightarrow \omega^2 = \frac{k}{m}$$

Angular frequency tells you the number of radians traversed per second, and it corresponds to the angular velocity for an object traveling in uniform circular motion. Be careful, however. Though angular frequency and angular velocity use the same symbol and are closely related in terms of definition, they are not synonymous.

Angular frequency is, as you might expect, closely related to frequency and period. Recall that frequency tells you the number of cycles or revolutions per second, in units of seconds^{-1}, or hertz (Hz), while period tells you the time for one complete cycle or revolution, in seconds. Noting that one complete revolution of a circle is 2π radians, angular frequency can be easily derived as:

$$\omega = 2\pi f \xrightarrow{T=\frac{1}{f}} \omega = \frac{2\pi}{T}$$

Putting this all together, we can refine the derivation of the simple harmonic motion differential equation to deliver a more general and useful version of the solution in which the position of the object as a function of time is given in terms of the angular frequency of the object.

$$\frac{d^2x}{dt^2} + \frac{k}{m}x = 0 \xrightarrow[\omega^2=\frac{k}{m}]{\omega=\sqrt{\frac{k}{m}}} \frac{d^2x}{dt^2} + \omega^2 x = 0 \rightarrow x(t) = A\cos(\omega t)$$

> **Author's Note**: The solution to this equation could also be written in terms of the sine function, as the sine and cosine function follow the same shape, but with a phase shift of $\phi = \frac{\pi}{2}$ radians, which may be accounted for in the solution as $x(t) = A\sin(\omega t + \phi)$.

A more general form describing the motion of an object undergoing SHM can be written as $x(t) = A\cos(\omega t + \phi)$, where the symbol phi ($\phi$) refers to the **phase angle**, or starting point, of a sine or cosine curve. Use the cosine curve if the motion starts at maximum amplitude. Use the sine curve if the motion starts at x=0. If the motion being described starts somewhere between maximum amplitude and equilibrium (x=0), you'll have to add in a phase angle component.

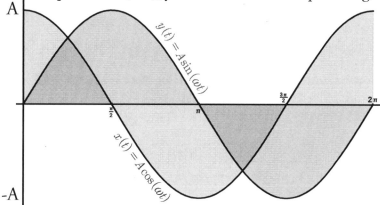

Having found the position of an object undergoing simple harmonic motion as a function of time, you can determine the velocity of the object by taking the derivative with respect to time.

$$x(t) = A\cos(\omega t) \rightarrow v = \frac{dx}{dt} = \frac{d}{dt}(A\cos(\omega t)) = -\omega A\sin(\omega t)$$

Because the range of the sine function is -1 to 1, the magnitude of the maximum velocity of the object must be ωA.

In similar fashion, you can determine the acceleration of the object by taking the derivative of velocity with respect to time, which is the second derivative of position with respect to time.

$$a = \frac{dv}{dt} = \frac{d^2x}{dt^2} = \frac{d}{dt}(-\omega A\sin(\omega t)) = -\omega^2 A\cos(\omega t)$$

The magnitude of the maximum acceleration of the object must be $\omega^2 A$ since the range of the cosine function is -1 to 1.

8.1 Q: A system oscillating in simple harmonic motion is created by releasing an object from a maximum displacement of 0.2 meters. The object makes 60 complete oscillations in one minute.
A) Determine the object's angular frequency.
B) What is the object's position at time t=10s?
C) At what time is the object at x=0.1m?

8.1 A: A) $\omega = 2\pi f = 2\pi(1\ Hz) = 2\pi\ ^{rad}/_s$

B) $x = A\cos(\omega t) \xrightarrow[A=0.2\,m]{\omega=2\pi^{rad}/_s} x = 0.2\cos(2\pi t) \xrightarrow{t=10s} x = 0.2\cos(2\pi(10)) = 0.2\ m$

C) $x = 0.2\cos(2\pi t) \rightarrow \cos(2\pi t) = \frac{x}{0.2} \rightarrow 2\pi t = \cos^{-1}\left(\frac{x}{0.2}\right) \rightarrow t = \frac{\cos^{-1}\left(\frac{x}{0.2}\right)}{2\pi} \rightarrow$

$t = \frac{\cos^{-1}\left(\frac{0.1}{0.2}\right)}{2\pi} = 0.167s$

8.2 Q: Which of the following statements about an object oscillating in simple harmonic motion about its equilibrium point is false?
A) The displacement is directly related to the acceleration.
B) The acceleration and velocity vectors always point in the same direction.
C) The acceleration vector is always toward the equilibrium point.
D) The acceleration and displacement vectors always point in opposite directions.

8.2 A: (B) The acceleration and velocity vectors always point in the same direction.

8.3 Q: The position of a particle is described by the equation $x(t) = (8\,cm)\cos(2\pi t)$.
A) What is the angular frequency?
B) What is the frequency?
C) What is the period?
D) When does the particle first reach its equilibrium position after time t=0?
E) What is the maximum speed?
F) What is the maximum acceleration?

8.3 A: A) $\omega = 2\pi$

B) $f = \dfrac{\omega}{2\pi} = \dfrac{2\pi}{2\pi} = 1\,Hz$

C) $T = \dfrac{1}{f} = \dfrac{1}{\frac{\omega}{2\pi}} = \dfrac{2\pi}{\omega} = \dfrac{2\pi}{2\pi} = 1s$

D) $x(t) = 0 \to 2\pi t = \dfrac{\pi}{2} \to t = \dfrac{\pi/2}{2\pi} = \dfrac{\pi}{2}\dfrac{1}{2\pi} = 0.25s$

E) $x(t) = 0.08\cos(2\pi t) \to v = \dfrac{dx}{dt} = \dfrac{d}{dt}(0.08\cos(2\pi t)) = 0.08\dfrac{d}{dt}\cos(2\pi t) =$
$(0.08)(-\sin(2\pi t))(2\pi) = -0.503\sin(2\pi t) \to v_{max} = 0.503\,{}^m\!/_s$

F) $a(t) = \dfrac{dv}{dt} = \dfrac{d}{dt}(-0.503\sin(2\pi t)) = -0.503\dfrac{d}{dt}\sin(2\pi t) =$
$(-0.503)(-\cos(2\pi t))(2\pi) = 3.16\cos(2\pi t) \to a_{max} = 3.16\,{}^m\!/_{s^2}$

8.4 Q: The engine on a toy train moves at constant speed in a circular path of radius 0.7 m, making one complete revolution in 8 seconds.

R=0.7 m

A) What is the speed of the engine?
B) What is the frequency with which the engine circles the track?
C) What is the angular velocity of the engine?
D) What is the acceleration of the engine?
E) What is the angular frequency of the engine?
F) Determine an equation for the engine's x-position as a function of time, assuming the engine is on the positive x-axis (where the head of the radius arrow is located in the diagram above) at time t=0.
G) Determine an equation for the engine's velocity in the x-direction as a function of time. What is its maximum velocity in the x-direction?
H) Determine an equation for the engine's acceleration in the x-direction as a function of time. What is its maximum acceleration in the x-direction?
I) Compare your answers to A&G, C&E, D&H. Step back and take a moment to convince yourself of why this makes sense.

8.4 A:

A) $v = \dfrac{C}{T} = \dfrac{2\pi r}{T} = \dfrac{2\pi\,(0.7m)}{8s} = 0.550\,{}^{m}\!/_{s}$

B) $f = \dfrac{1}{T} = \dfrac{1}{8s} = 0.125\,Hz$

C) $\omega = \dfrac{v}{R} = \dfrac{0.550\,{}^{m}\!/_{s}}{0.7m} = 0.785\,{}^{rad}\!/_{s}$

D) $a_{c} = \dfrac{v^{2}}{R} = \dfrac{\left(0.550\,{}^{m}\!/_{s}\right)^{2}}{0.7m} = 0.432\,{}^{m}\!/_{s^{2}}$

E) $\omega = 2\pi f = 2\pi\,(0.125\,Hz) = 0.785\,{}^{rad}\!/_{s}$

F) $x(t) = A\cos(\omega t) = 0.7\cos(0.785t)$

G) $v(t) = \dfrac{dx}{dt} = \dfrac{d}{dt}(0.7\cos(0.785t)) = -0.550\sin(0.785t) \rightarrow v_{max} = 0.550\,{}^{m}\!/_{s}$

H) $a(t) = \dfrac{dv}{dt} = \dfrac{d}{dt}(-0.550\sin(0.785t)) = 0.432\cos(0.785t) \rightarrow a_{max} = 0.432\,{}^{m}\!/_{s^{2}}$

I) They are the same.

8.5 Q: Which of the following are most likely to result in simple harmonic motion? Select two answers.
A) A hole is drilled through one end of a meter stick, which is hung vertically from a frictionless axle. The bottom of the meter stick is displaced 12 degrees from vertical and released.
B) A hole is drilled through one end of a meter stick, which is hung vertically from a rough axle. The bottom of the meter stick is displaced 12 degrees and released. Every time the meter stick swings back and forth the axle squeaks.
C) A block is hung vertically from a linear spring. The opposite end of the spring is attached to a stationary point. The entire apparatus is placed in deep space. The block is displaced 4 cm from equilibrium and released.
D) A block is placed on a frictionless surface and attached to a non-linear spring. The opposite end of the spring is attached to a wall. The block is displaced 2 cm from equilibrium and released.

8.5 A: (A) and (C). Simple harmonic motion required a linear restoring force. The real pendulum described in answer (A) can be readily approximated using simple harmonic motion using the small angle approximation (described later in this chapter). Answer (B) can be eliminated due to the loss of energy from the rough axle, which will result in damping. Answer (C) describes true simple harmonic motion in a frictionless environment with a spring-block oscillator. Answer (D) can be eliminated due to the non-linear spring.

Spring-Block Oscillators

A popular demonstration vehicle for simple harmonic motion is the spring-block oscillator. The horizontal spring-block oscillator consists of a block of mass m sitting on a frictionless surface, attached to a vertical wall by a spring of spring constant k, as shown in the diagram below.

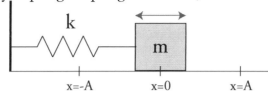

The block is then displaced an amount A from its equilibrium position and allowed to oscillate back and forth. As the block sits on a frictionless surface, in the ideal scenario the block would continue its periodic motion indefinitely. It is this situation that was used initially to derive the formulas describing simple harmonic motion.

8.6 Q: A 5-kg block is attached to a 2000 N/m spring as shown and displaced a distance of 8 cm from its equilibrium position before being released.

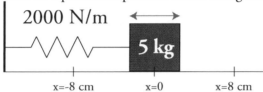

Determine the period of oscillation, the frequency, and the angular frequency for the block.

8.6 A:
$$T_s = 2\pi\sqrt{\frac{m}{k}} = 2\pi\sqrt{\frac{5\ kg}{2000\ ^N/_m}} = 0.314\ s$$

$$f = \frac{1}{T} = \frac{1}{0.314s} = 3.18\ Hz$$

$$\omega = 2\pi f = 2\pi\,(3.18\ Hz) = 20\ ^{rad}/_s$$

8.7 Q: Rank the following horizontal spring-block oscillators resting on frictionless surfaces in terms of their period, from longest to shortest.

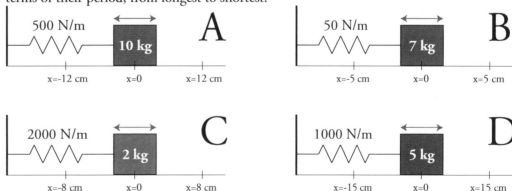

8.7 A: B, A, D, C

It's also interesting to look at the energy of the spring-block oscillator while its undergoing simple harmonic motion. Because the surface is frictionless, the total energy of the system remains constant. However, there is a continual transfer of kinetic energy into elastic potential energy and back.

When the block is at its equilibrium position, there is no elastic potential energy stored in the spring; therefore, all of the energy of the block is kinetic. The block has achieved its maximum speed. At this position, there is also no net force on the block, so the block's acceleration is zero.

When the block is at its maximum amplitude position, all of its energy is stored in the spring as elastic potential energy. For an instant its kinetic energy is zero, therefore its velocity is zero. Further, at this position, the spring exhibits a maximum force on the block, providing the maximum acceleration. Let's take a look at this graphically by examining a spring-block oscillator at various points in its periodic path.

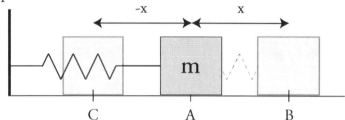

Graphs of Related Physical Quantities

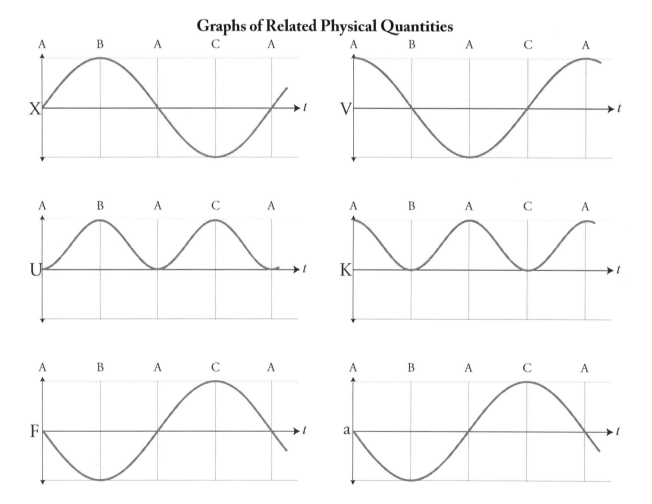

	A	B	C
Displacement (x)	0	X	-X
Velocity (v)	max	0	0
Potential Energy (U)	0	max	max
Kinetic Energy (K)	max	0	0
Force (F)	0	-max	max
Acceleration (a)	0	-max	max

Of course, through the entire time interval, the total mechanical energy of the spring-block oscillator remains constant.

In reality, some energy is typically lost to friction or other non-conservative forces. Over time, the amplitude of the oscillations steadily decrease. This is known as **damping**, or damped harmonic motion.

8.8 Q: A linear spring with spring constant 40 N/m is attached to a fixed surface, and a block of mass 0.25 kg is attached to the end of the spring, sitting on a frictionless surface. The block is now displaced 15 centimeters and released at time t=0.

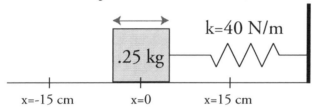

A) Draw a free body diagram for the block at t=0.
B) Determine the period of the spring-block oscillator.
C) Determine the speed of the block at position x=0.
D) Write an expression for the displacement of the block as a function of time.
E) Create the following plots. Explicitly label axes with units as well as any intercepts, asymptotes, maxima, or minima with numerical values or algebraic expressions, as appropriate.
  1) Displacement vs. Time
  2) Speed vs. Time
  3) Acceleration vs. Time
  4) Kinetic Energy and Elastic Potential Energy vs. Time

8.8 A: A)

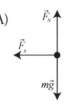

B) $T_s = 2\pi\sqrt{\dfrac{M}{k}} = 2\pi\sqrt{\dfrac{0.25\ kg}{40\ ^N/_m}} = 0.497\ s$

C) $U_s = K \rightarrow \frac{1}{2}kx^2 = \frac{1}{2}mv^2 \rightarrow v = \sqrt{\dfrac{kx^2}{M}} = \sqrt{\dfrac{(40\ ^N/_m)(0.15m)^2}{0.25\ kg}} = 1.90\ ^m/_s$

D) $x(t) = A\cos(\omega t) \rightarrow x(t) = 0.15\cos(12.6t)$

E)

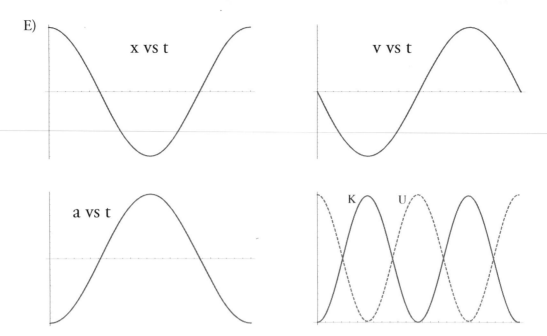

x vs t

v vs t

a vs t

K   U

8.9 Q: A disk with mass M, radius R, and rotational inertia $\frac{1}{2}MR^2$ is attached to a horizontal spring which has a spring constant of k as shown in the diagram.

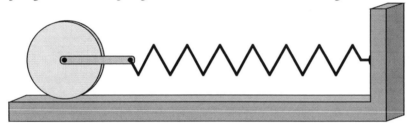

When the spring is stretched by a distance x and then released from rest, the disk rolls without slipping while the spring is attached to the frictionless axle within the center of the disk.

A) Calculate the maximum translational velocity of the disk in terms of M, R, x, and k.

B) What would happen to the period of this motion if the spring constant of the spring increased? Justify your answer.

C) What would happen to the period of this motion if the surface was now frictionless and the disk was not allowed to roll? Justify your answer.

8.9 A: A) Use conservation of energy to find the maximum velocity:

$$\tfrac{1}{2}kx^2 = \tfrac{1}{2}mv^2 + \tfrac{1}{2}I\omega^2 \xrightarrow[\omega=\frac{v}{R}]{I=\frac{1}{2}MR^2} v_{max} = x\sqrt{\frac{2k}{3M}}$$

B) If the spring constant k increased, the period of the motion would decrease. When you increase the k value of a spring the period of the spring decreases. This is due to the velocity of the object attached to the spring increasing. You can also look at the answer in part (A) and see that the velocity increases with a larger k, which means the time to travel one full oscillations will decrease.

C) There is a decrease in the period if the system is now on a frictionless surface. The disk will no longer roll and no energy is being put into rolling the disk. The total energy of the system will stay the same but more energy is now available for the translational kinetic energy (assuming you stretch the spring the same distance x). This leads to a higher velocity, which leads to a decrease in the period.

8.10 Q: A 5-kg mass attached to a linear spring undergoes simple harmonic motion along a frictionless tabletop with an amplitude of 0.35 m and a frequency of 0.67 Hz.
A) Explain what the two criteria are for simple harmonic motion.
B) Calculate the value of the spring constant.
C) What is the ratio of the mass's acceleration when it is at half its amplitude to its acceleration when it is at full amplitude?
D) Sketch a graph of the mass's acceleration as a function of velocity for half of a cycle. Start when the mass is at its greatest positive amplitude. Mark the initial acceleration as a_0.

8.10 A: A) The restoring force must be proportional to the displacement, and the restoring force must oppose the motion of the mass.

B) $f = \dfrac{1}{2\pi}\sqrt{\dfrac{k}{m}} \rightarrow k = (2\pi f)^2 m = (2\pi (0.67Hz))^2 (5kg) = 89\,{}^N\!/_m$

C) The acceleration is proportional to the force, which is proportional to the displacement according to Hooke's Law, therefore the ratio of $a/a_0 = 0.5$.
D) When x is at a maximum, v is 0 and a is at a maximum. Acceleration is negative when displacement is positive. Velocity is positive for the entire half cycle. Velocity is proportional to the sine function and acceleration is proportional to the cosine function (out of phase), resulting in an elliptical plot.

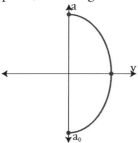

Vertical Spring-Block Oscillators

Spring-block oscillators can also be set up vertically as shown in the diagram.

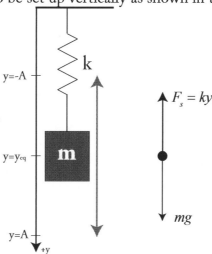

Start your analysis by drawing a Free Body Diagram for the block, noting that gravity pulls the mass down, while the force of the spring provides the upward force. Call down the positive y-direction. At its equilibrium position, $y=y_{eq}$. You can then write a Newton's 2nd Law Equation for the block and solve for y_{eq}.

$$F_{net_y} = mg - ky = ma_y \xrightarrow{\;equilibrium\;} mg - ky = 0 \rightarrow y_{eq} = \frac{mg}{k}$$

Once the system has settled at equilibrium, you can displace the mass by pulling it some amount to either +A or lifting it an amount -A. The new system can be analyzed as follows:

$$F_{net_y} = mg - k(y_{eq} + A) = mg - ky_{eq} - kA \xrightarrow{\;mg - ky_{eq} = 0\;} F_{net_y} = -kA$$

This is the same analysis you would do for a horizontal spring system with spring constant k displaced an amount A from its equilibrium position. This means, in short, that to analyze a vertical spring system, all you do is find the new equilibrium position of the system, taking into account the effect of gravity, then treat it as a system with only the spring force to deal with, oscillating around the new equilibrium point. No need to continue to deal with the force of gravity!

8.11 Q: A 2-kg block attached to an unstretched spring of spring constant k=200 N/m as shown in the diagram at right is released from rest.
A) Determine the period of the block's oscillation.
B) What is the maximum displacement of the block from its equilibrium while undergoing simple harmonic motion?

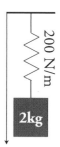

8.11 A: A) $T_s = 2\pi\sqrt{\dfrac{m}{k}} = 2\pi\sqrt{\dfrac{2\,kg}{200\,{}^{N}\!/_{m}}} = 0.63\,s$

B) The gravitational potential energy at the block's starting point, $mg\Delta y$, must equal the elastic potential energy stored in the spring at its lowest point. Use this to solve for Δy.

$$U_g = U_s \rightarrow mg\Delta y = \tfrac{1}{2}k\Delta y^2 \rightarrow \Delta y = \frac{2mg}{k}$$

If Δy is the total displacement of the block from its highest point to its lowest point, the maximum displacement of the block from its equilibrium point, A, must be half of Δy.

$$A = \frac{\Delta y}{2} = \frac{mg}{k} = \frac{(2\,kg)(9.8\,{}^{m}\!/_{s^2})}{200\,{}^{N}\!/_{m}} = 0.1\,m$$

8.12 Q: A 5-kg block is attached to a vertical spring (k=500 N/m). After the block comes to rest, it is pulled down 3 cm and released.
A) What is the period of oscillation?
B) What is the maximum displacement of the spring from its initial unstrained position?

8.12 A: A) $T_s = 2\pi\sqrt{\dfrac{m}{k}} = 2\pi\sqrt{\dfrac{5\,kg}{500\,{}^{N}\!/_{m}}} = 0.63\,s$

B) First determine the displacement of the spring when the block is hanging and at rest.

$$F_{net_y} = 0 \rightarrow kd = mg \rightarrow d = \frac{mg}{k} = \frac{(5\,kg)(9.8\,{}^{m}\!/_{s^2})}{500\,{}^{N}\!/_{m}} = 0.1\,m$$

Then the block is pulled down 3 cm, so the maximum displacement must be the displacement while the block is at rest in addition to the 3 cm the block is pulled down.

$$y_{max} = 0.1\,m + 0.03\,m = 0.13\,m$$

8.13 Q: A spring of spring constant k is hung vertically from a fixed surface, and a block of mass M is attached to the bottom of the spring. The mass is released and the system is allowed to come to equilibrium as shown in the diagram at right.

A) Derive an expression for th equilibrium position of the mass.

B) The spring is now pulled downward and displaced an amount A.

C) At time t=0, the spring is released. Derive an expression for the period of the spring-block oscillator.

D) Describe an experimental procedure you could use to verify your derivation of the period. Include all equipment required.

E) How would you analyze the data to determine whether the experimental data verifies your derivation? What evidence from the analysis would be used to make the determination?

8.13 A: A) $y_{eq} = \dfrac{Mg}{k}$

B) $U_s = \dfrac{1}{2}kA^2$

C) $T_s = 2\pi\sqrt{\dfrac{M}{k}}$

D) One method could involve utilizing a stopwatch to measure the time it takes for 10 oscillations, and dividing that by 10 to obtain the experimental period for the given mass. A more detailed analysis could include repeating this for a variety of masses. Equipment required would include a stopwatch and hanging masses. A variety of alternate acceptable answers exist, which could include, but are not limited to, electronic measuring devices such as photogates.

E) One method would involve plotting the period vs. the square root of the mass. This plot should be linear, with the slope equal to 2π divided by $\sqrt{k}$. Including an uncertainty analysis of the data, if the calculated value for the spring constant from the slope of the graph matches the spring constant of the spring (within its uncertainty), it would be reasonable to conclude that the derivation is correct.

Spring Combinations

It is also possible to attach more than one spring to an object. In these cases, analyses can be simplified considerably by treating the combination of springs as a single spring with an equivalent spring constant.

Springs in Parallel

For springs in parallel, calculate an equivalent spring constant for the system by starting with Hooke's Law, recognizing that displacement is the same for both springs.

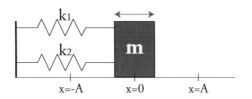

$$F = k_1 x + k_2 x = (k_1 + k_2) x = k_{eq} x \rightarrow k_{eq} = k_1 + k_2$$

Springs in Series

For springs in series, you will again calculate an equivalent spring constant for the system, beginning the analysis by realizing the force on each spring must be the same according to Newton's 3rd Law of Motion.

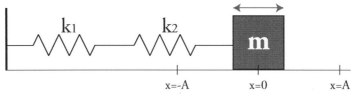

$$F = k_1 x_1 = k_2 x_2 \rightarrow x_1 = \frac{k_2}{k_1} x_2$$

Next, recognizing that the total displacement is equal to the sum of the displacement of the springs, you can combine the equations and solve for an equivalent spring constant.

$$F = k_{eq}(x_1 + x_2) \xrightarrow{x_1 = \frac{k_2}{k_1} x_2} F = k_{eq}\left(\frac{k_2}{k_1} x_2 + x_2\right) \xrightarrow{F = k_2 x_2} k_2 x_2 = k_{eq} k_2 \left(\frac{k_2}{k_1} + 1\right) \rightarrow \frac{1}{k_{eq}} = \frac{1}{k_1} + \frac{1}{k_2}$$

8.14 Q: Rank the spring block oscillators in the diagram below from highest to lowest in terms of:
I) equivalent spring constant
II) period of oscillation

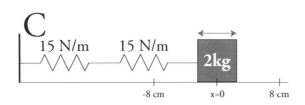

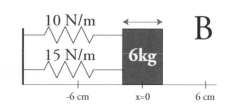

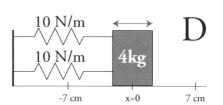

8.14 A: I) B, D, C, A
II) A, C, B, D

The Pendulum

Ideal pendulums provide another demonstration vehicle for simple harmonic motion. Consider a mass M attached to a light string that swings without friction about the vertical equilibrium position as shown below left. As the mass travels along its path, energy is continuously transferred between gravitational potential energy and kinetic energy. The restoring force in the case of the ideal pendulum is provided by gravity.

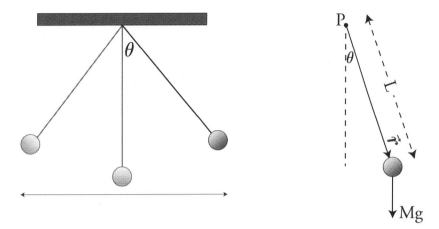

You can find the period of a simple (ideal) pendulum by starting with the definition of torque. Assume a point mass M hangs from pivot point P by a light string of length L, displaced from its vertical equilibrium position by an angular displacement θ as shown above right.

$$\vec{\tau}_P = \vec{r}_P \times \vec{F} = \vec{r}_P \times Mg \rightarrow |\vec{\tau}_P| = MgL\sin\theta \xrightarrow{\tau_{net}=I\alpha} -MgL\sin\theta = I_P\alpha$$

Note that the angular acceleration as shown in the diagram in the upper right will be lockwise, while we defined theta, according to the diagram, as counter-clockwise. We account for this with the negative sign in front of $MgL\sin\theta$.

Next, we employ a small-angle approximation. The small-angle approximation for the sine function states that for small angles, the value of $\sin\theta$ is very close to θ itself (within about one percent error for angles under roughly 15 degrees). Since we'll be employing this approximation, our solution for the period will only be usable for situations in which the angular displacement of the mass is small.

$$-MgL\sin\theta = I_P\alpha \xrightarrow[\theta<15°]{\sin\theta\approx\theta} -MgL\theta = I_P\alpha \xrightarrow{\alpha=\frac{d^2\theta}{dt^2}} -MgL\theta = I_P\frac{d^2\theta}{dt^2} \rightarrow \frac{d^2\theta}{dt^2} + \frac{MgL}{I_P}\theta = 0$$

This is a differential equation which fits the form of simple harmonic motion. Further, the argument in front of theta must be the square of the angular frequency.

$$\omega^2 = \frac{MgL}{I_P} \rightarrow \omega = \sqrt{\frac{MgL}{I_P}}$$

The solution to the differential equation, of course, is $\theta = A\cos(\omega t)$. Further, the moment of inertia of a point mass M a distance L from its axis of rotation is easily determined as ML^2, allowing us to solve for the common form of the period of a simple pendulum.

$$\omega = \sqrt{\frac{MgL}{I_P}} \xrightarrow{I_P=ML^2} \omega = \sqrt{\frac{MgL}{ML^2}} = \sqrt{\frac{g}{L}} \xrightarrow{T=\frac{2\pi}{\omega}} T = \frac{2\pi}{\sqrt{\frac{g}{L}}} \rightarrow T = 2\pi\sqrt{\frac{L}{g}}$$

8.15 Q: A grandfather clock is designed such that each swing (or half-period) of the pendulum takes one second. How long is the pendulum in a grandfather clock?

8.15 A: $T_P = 2\pi\sqrt{\dfrac{L}{g}} \rightarrow L = \dfrac{gT^2}{4\pi^2} = \dfrac{(9.8\,{}^m\!/_{s^2})(2s)^2}{4\pi^2} = 1m$

8.16 Q: What is the period of a grandfather clock on the moon, where the acceleration due to gravity on the surface is roughly one-sixth that of Earth?

8.16 A: $T_P = 2\pi\sqrt{\dfrac{L}{g}} = 2\pi\sqrt{\dfrac{1m}{\frac{1}{6}(9.8\,{}^m\!/_{s^2})}} = 4.9\,s$

8.17 Q: Rank the following pendulums of uniform mass density from highest to lowest frequency.

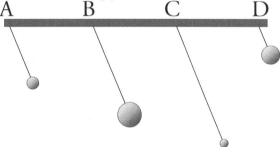

8.17 A: D, A, B, C

Closely observing the geometric representation of the pendulum, you can construct a diagram detailing the energy and forces acting on the pendulum at various points in its parabolic path.

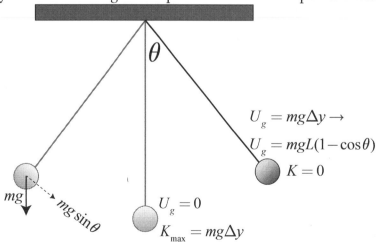

Chapter 8: Oscillations

At the highest point, as shown on the left, the mass is pulled back toward its equilibrium position by gravity. Specifically, the component of gravity along the mass's path, mgsin θ. At the equilibrium position, the gravitational potential energy is at a minimum, and the kinetic energy of the mass is at a maximum. At the highest point, as shown on the right, all the energy is gravitational potential energy again. Using the law of conservation of energy, solving for the maximum velocity of the mass at its lowest position is quite straightforward.

$$K = U_g \rightarrow \tfrac{1}{2}mv^2 = mgL(1-\cos\theta) \rightarrow v = \sqrt{2gL(1-\cos\theta)}$$

Further, from this same analysis you can create a graph of kinetic energy, gravitational potential energy, and total energy as a position of the mass along the x-axis.

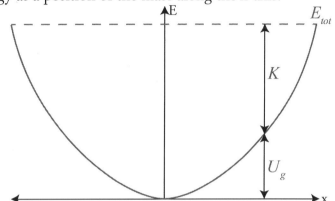

The magnitude of the energy below the parabolic line represents the gravitational potential energy of the mass, while above the line the kinetic energy is represented. The total mechanical energy remains constant throughout the entire path of the pendulum.

8.18 Q: The period of an ideal pendulum is T. If the mass of the pendulum is tripled while its length is quadrupled, what is the new period of the pendulum?
(A) 0.5 T
(B) T
(C) 2T
(D) 4T

8.18 A: (C) 2T

8.19 Q: Which of the following are true for an ideal pendulum consisting of a mass oscillating back and forth on a light string? (Choose all that apply.)
(A) The kinetic energy is always equal to the potential energy.
(B) The maximum force on the pendulum occurs when the pendulum has its maximum kinetic energy.
(C) The tangential acceleration of the pendulum is zero when the mass is at its lowest point.
(D) The angular acceleration of the mass remains constant.

8.19 A: (C) The tangential acceleration of the pendulum is zero when the mass is at its lowest point.

8.20 Q: A pendulum of length 20 cm and mass 1 kg is displaced an angle of 10 degrees from the vertical. What is the maximum speed of the pendulum?

8.20 A: $v = \sqrt{2gL\left(1 - \cos\theta\right)} = \sqrt{2\left(9.8\,{}^{m}\!/_{s^2}\right)\left(0.2m\right)\left(1 - \cos 10°\right)} = 0.24\,{}^{m}\!/_{s}$

8.21 Q: A pendulum of length 0.5 m and mass 5 kg is displaced an angle of 14 degrees from the vertical. What is the speed of the pendulum when its angle from the vertical is 7 degrees?

8.21 A: $U_{g_{top}} = U_{g_{bottom}} + K_{bottom} \rightarrow mgL\left(1 - \cos 14°\right) = mgL\left(1 - \cos 7°\right) + \tfrac{1}{2}mv^2 \rightarrow$
$v = \sqrt{2gL\left(\cos 7° - \cos 14°\right)} = 0.47\,{}^{m}\!/_{s}$

The Physical Pendulum

So far, we've investigated the case of a simple, or ideal, pendulum, in which all the mass is concentrated in a point particle, and the rod or string connecting the mass to the pivot point is considered massless. Now, we'll explore the period of a physical pendulum, in which mass exists in locations other than just a point mass. A great example of a physical pendulum is a meter stick. Consider how a pendulum can be made from a meter stick by drilling a hole at some distance from one end, displacing the ruler some small angle from the vertical, and allowing it to oscillate back and forth.

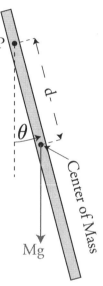

We'll begin our derivation by making a diagram, where P represents the pivot point, d is the distance from the pivot point to the center of mass of the physical pendulum, and $\vec{r}_P$ is the position vector from the pivot point to the center of mass. The length of the physical pendulum is L, and the mass of the pendulum, M, is uniformly distributed along its length.

Just like the analysis of the ideal pendulum, the analysis of the physical pendulum begins with the definition of the net torque about reference point P.

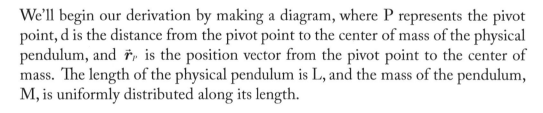

Again employing the small angle approximation for angles less than approximately 15 degrees, you can develop a differential equation fitting the form of simple harmonic motion.

$$-Mgd\sin\theta = I_P\alpha \xrightarrow[\theta < 15°]{\sin\theta \approx \theta} -Mgd\theta = I_P\alpha \xrightarrow{\alpha = \frac{d^2\theta}{dt^2}} -Mgd\theta = I_P\frac{d^2\theta}{dt^2} \rightarrow \frac{d^2\theta}{dt^2} + \frac{Mgd}{I_P}\theta = 0$$

This form of the solution indicates that the square of the angular frequency (ω^2) must be equal to the Mgd/I_P, leading to the general solution for the period of the physical pendulum.

$$T = \frac{2\pi}{\omega} \xrightarrow{\omega = \sqrt{\frac{Mgd}{I_P}}} T = 2\pi\sqrt{\frac{I_P}{Mgd}}$$

Progressing further through the derivation, you can find the moment of inertia of the physical pendulum about point P using the parallel axis theorem.

$$I_P = I_{CoM} + Md^2 \xrightarrow{\ I_{CoM} = \frac{ML^2}{12}\ } I_P = \frac{ML^2}{12} + Md^2$$

Substitute in this value for the moment of inertia about point P into the general solution for the period of a physical pendulum to obtain the final expression for the period.

$$T = 2\pi\sqrt{\frac{I_P}{Mgd}} \xrightarrow{\ I_P = \frac{ML^2}{12} + Md^2\ } T = 2\pi\sqrt{\frac{\frac{ML^2}{12} + Md^2}{Mgd}} = 2\pi\sqrt{\frac{\frac{L^2}{12} + d^2}{gd}}$$

Notice that there is no mass dependence on the period of the physical pendulum.

8.22 Q: A small hole is drilled into a meter stick 20 cm from the end of the stick to serve as a pivot point. The meter stick is placed on an axle through the hole and displaced a small amount and allowed to oscillate back and forth. How long does it take the meter stick to complete 60 complete oscillations?

8.22 A: Begin by noting that a hole drilled 20 cm from the end of the meter stick indicates the pivot point will be located 30 cm from the center of mass of the meter stick.

$$T = 2\pi\sqrt{\frac{\frac{L^2}{12} + d^2}{gd}} \xrightarrow[d=0.3\,m]{L=1\,m} T = 2\pi\sqrt{\frac{\frac{1^2}{12} + (0.3)^2}{(9.8)(0.3)}} = 1.53\,s$$

Find the total time for 60 oscillations.
$$t = 60T = 60\,(1.53\,s) = 91.5\,s$$

8.23 Q: A thin disc of radius 30 cm and mass 2 kg is pivoted about a tiny hole at its very edge. The disk is rotated 8 degrees from the vertical and allowed to oscillate. Determine the period of the disc.

8.23 A: First find the moment of inertia of the disc rotated about a point on its edge using the parallel axis theorem.
$$I = I_{CoM} + Md^2 = \tfrac{1}{2}MR^2 + MR^2 = \tfrac{3}{2}MR^2$$

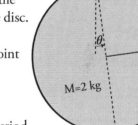

Next, solve for the period using the general solution for the period of a physical pendulum.

$$T = 2\pi\sqrt{\frac{I_P}{Mgd}} \xrightarrow[d=R]{I_P = \frac{3}{2}MR^2} T = 2\pi\sqrt{\frac{\frac{3}{2}MR^2}{MgR}} = 2\pi\sqrt{\frac{3R}{2g}} = 2\pi\sqrt{\frac{3(0.3)}{2(9.8)}} = 1.35\,s$$

8.24 Q: On a flat surface in Toledo, Ohio, a simple pendulum of length 0.58 m and mass 0.34 kg is pulled back an angle of 45 degrees and released.
A) Determine the theoretical period of the pendulum.
B) When the period is measured (with a photogate, for 20 oscillations) it is found that the experimental period is five percent higher than expected. Repeated measurements consistently give similar values. What is the most likely explanation for this systematic error?
C) What would be the period of this pendulum if it was in a free-fall environment?

8.24 A: A) $T = 2\pi\sqrt{\dfrac{L}{g}} = 2\pi\sqrt{\dfrac{0.58\,m}{9.8\,\text{\tiny m}/s^2}} = 1.53\,s$

B) The most likely explanation is the theoretical equation is under-calculating the period. The equation used in part (A) is based on the small angle approximation and in the lab situation the angle is 45 degrees, which is too large an angle to apply the small angle approximation. Consequently, a large angle gives a slightly greater period than expected.
C) There would be no period (or $T = \infty$).

8.25 Q: A group of students want to design an experiment where they test the period of a simple pendulum as a function of the acceleration due to gravity, g. Part of their motivation is to verify the equation of the period of a simple pendulum: $T = 2\pi\sqrt{\dfrac{L}{g}}$. They propose to bring their apparatus on NASA's Zero-g airplane that can create a range of values of g for short periods of time. Their proposal calls for three trials each for two situations where $g < 9.8$ m/s^2 (where everything feels lighter than normal), g=9.8 m/s^2, and two situations where $g > 9.8$ m/s^2 (where everything feels heavier than normal).

A) If the Zero-g airplane moves in a continuous series of up-and-down parabolas, as shown below, identify at which position(s) in its trajectory someone would feel heavier than normal.

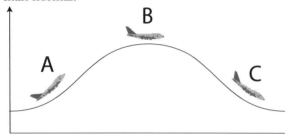

B) If the simple pendulum the students bring consists of a 1.0-kg sphere with a string attached, discuss two other important factors that must be controlled throughout the experiment and how they will be controlled.
C) Since the students have a limited number of trials for each value of g, they need a method of minimizing uncertainty in the measurement of the period of the pendulum. Discuss a method for doing so. Be sure to list necessary equipment and how it will be applied.
D) Assuming the students carry out the experiment and gather valid data, they want to graph their data. If they plot the variable T^2 on the y-axis, what variable should they plot on the x-axis so they get a straight line of best fit?
E) When they draw the line of best fit, the students calculate a slope of 9.8 m. What is the length of their pendulum?

8.25 A: A) Heavier than normal at positions A and C.
B) Length of string can be controlled by securing one end to the sphere and the other end to a clamp system. The angle (amplitude) can be controlled by pulling back a predetermined (small) distance. Perhaps have a marker to pull back to each time.
C) Let the pendulum swing back and forth five to 10 times and time with a stopwatch or photogate system.
D) Plot T^2 vs 1/g (accept T^2 vs k/g where k is some constant value such as $4\pi^2 L$ or L)
E) Assuming they plotted T^2 vs. 1/g, the slope is proportional to T^{2}g. Using the pendulum equation: $L = \dfrac{T^2 g}{4\pi^2} = \dfrac{slope}{4\pi^2} = 0.25\,m$

8.26 Q: Students are to conduct an experiment to investigate the relationship between the length of a pendulum and its period. Their procedure involves hanging a 0.5 kg mass on a string of varying length, L, setting it into oscillation, and measuring the period. The students conduct the experiment and obtain the following data.

Trial	1	2	3	4	5	6
Length (m)	0.25	0.50	0.75	1.0	1.5	2.0
Period (s)	1.0	1.4	1.7	2.0	2.5	2.8

A) Plot the period of the pendulum (T) as a function of the length of the string (L), and draw a best-fit curve. Label the axes as appropriate.

B) Plot a linear graph as a function of L. Use the empty boxes in the data table to record any calculated values you are graphing. Label the axes as appropriate.

Students are then given strings of various lengths, a hanging mass, a meter stick, a stopwatch, appropriate survival equipment, and are transported to the surface of Planet X. There, they are asked to determine the acceleration due to gravity on the surface of Planet X using just this equipment.

C) Describe an experimental procedure that the students could use to collect the necessary data as accurately as possible.

In order to determine the acceleration due to gravity, the students then create a plot of T vs. the square root of L, as shown below.

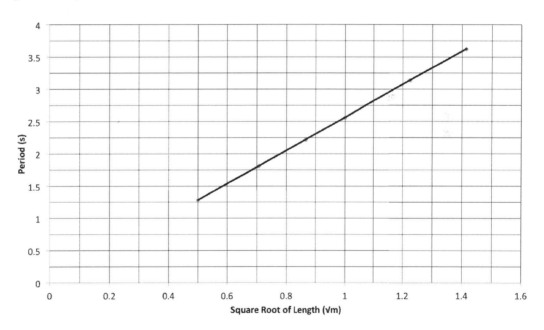

D) Using the graph, calculate the acceleration due to gravity on Planet X.

E) Following further analysis and experiments, a comparison of the students' experimental value for the acceleration due to gravity on Planet X is greater than the actual acceleration due to gravity on Planet X. Offer a reasonable explanation for this difference.

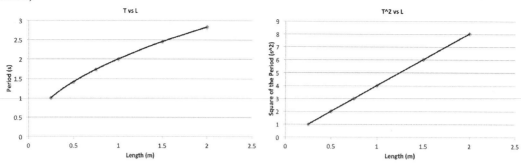

C) Hang the mass from the string and attach the top of the string to a fixed point. Pull the mass back a small angular displacement (approximately 10 degrees) and release. Use a stopwatch to time 20 oscillations, and divide the total time by 20 to obtain the period.

D) $g = \dfrac{4\pi^2}{slope^2} = \dfrac{4\pi^2}{2.57^2} = 6 \, ^m\!/_s$

E) This analysis assumes the string is massless. A real string has mass, which would lead to a smaller measured period of oscillation, and therefore a larger measured acceleration due to gravity compared with the actual acceleration due to gravity on the planet.

Resonance

Certain devices create strong oscillations at a single specific frequency. If another object, having the same "**natural frequency,**" is impacted by these oscillations, it may begin to increase the amplitude of the oscillations at this frequency. This phenomenon is known as **resonance**.

This resonance effect can be observed when you push someone on a swing. By pushing at the resonant frequency, the swing goes higher and higher! A dramatic demonstration of resonance involves an opera singer breaking a glass by singing a high pitch note. The singer creates a sound wave with a frequency equal to the natural frequency of the glass, causing the glass to oscillate at its natural, or resonant, frequency so energetically that it shatters.

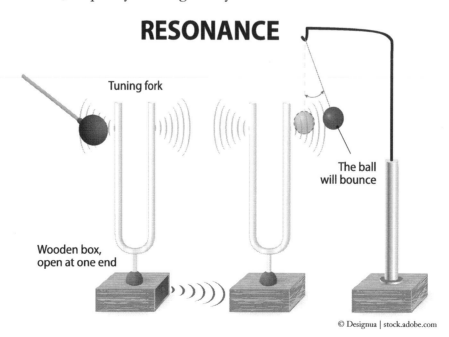

© Designua | stock.adobe.com

Chapter 9: Gravity

© Anterovium | stock.adobe.com

Objectives

1. Calculate the force a spherically symmetrical object exerts on another.

2. Determine the gravitational field strength at points inside and outside spherically symmetrical objects.

3. Analyze circular orbits to describe velocity, period, acceleration, kinetic energy, potential energy, and total energy.

4. State Keplers Three Laws of Planetary Motion.

5. Derive Kepler's Third Law of Planetary Motion for a circular orbit.

6. Define matter, mass, work and energy.

7. Recognize the intent and depth of this book as a companion resource to be used in conjunction with active learning practices such as hands-on exploration, discussion, debate, and deeper problem solving.

8. Determine velocity and radius for a circular or elliptical orbit using conservation of angular momentum.

9. Analyze the motion of an object that is projected straight up from a planet's surface.

G ravity is one of the strangest yet most common forces we encounter each and every day. We regularly model the effects of gravity with tremendous precision, yet understanding the "why" of gravity is an ongoing area of exploration for scientists to this day. In this chapter, we'll take a look how we can model the gravitational force of attraction between two objects, then apply this model to the motion of objects in space.

Newton's Law of Universal Gravitation

All objects that have mass attract each other with a gravitational force. Consider the case of two spherically symmetrical masses, m_1 and m_2. The center of the masses is separated by a distance r, and you can define a position vector $\vec{r}$ that runs from the center of the first mass to the center of the second mass.

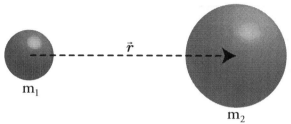

The magnitude of the gravitational force on the first mass is described by Newton's Law of Universal Gravitation:

$$\vec{F}_g = -G\frac{m_1 m_2}{r^2}\hat{r}$$

This law says that the force of gravity between two objects is proportional to each of the masses(m_1 and m_2) and inversely proportional to the square of the distance between their centers (r). The **universal gravitational constant**, G, is a "fudge factor," so to speak, included in the equation so that your answers come out in S.I. units. G is given as 6.67×10^{-11} N·m²/kg². The negative sign, coupled with the unit vector $\hat{r}$, tells you the direction of the force on m_1. Applying Newton's 3rd Law of Motion, the gravitational force on the second mass has the same magnitude, but is opposite in direction.

Let's look at this relationship in a bit more detail. Force is directly proportional to the masses of the two objects. Therefore, if either of the masses were doubled, the gravitational force would also double. In similar fashion, if the distance between the two objects, r, was doubled, the force of gravity would be quartered since the distance is squared in the denominator. This type of relationship is called an inverse square law, which describes many phenomena in the natural world.

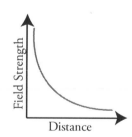

NOTE: The distance between the masses, r, is actually the distance between the center of masses of the objects. For large objects, such as the gravitational attraction between the Earth and the moon, you must determine the distance from the center of the Earth to the center of the moon, not their surfaces.

Previously, you learned that **inertial mass** is an object's resistance to being accelerated by a force, and more massive objects accelerate less than smaller objects given an identical force. You also learned that **gravitational mass** relates to the amount of gravitational force experienced by an object. Objects with larger gravitational mass experience a larger gravitational force. Thankfully, inertial mass and gravitational mass have been tested over and over again and are exactly the same.

Some hints for problem solving when dealing with Newton's Law of Universal Gravitation:
1. Substitute values in for variables at the very end of the problem only. The longer you can keep the formula in terms of variables, the fewer opportunities for mistakes.
2. Before using your calculator to find an answer, try to estimate the order of magnitude of the answer. Use this to check your final answer.
3. Once your calculations are complete, make sure your answer makes sense by comparing your answer to a known or similar quantity. If your answer doesn't make sense, check your work and verify your calculations.

9.1 Q: What is the magnitude of the gravitational force of attraction between two asteroids in space, each with a mass of 50,000 kg, separated by a distance of 3800 m?

9.1 A: $|\vec{F_g}| = G\dfrac{m_1 m_2}{r^2} = \left(6.67 \times 10^{-11} \; {}^{N \cdot m^2}/_{kg^2}\right) \dfrac{(50,000 \; kg)(50,000 \; kg)}{(3800 \; m)^2} = 1.15 \times 10^{-8} \; N$

9.2 Q: Determine the magnitude of the gravitational force exerted on the Earth by the sun given the mass of the Earth (6×10^{24} kg), the mass of the sun (2×10^{30} kg), and the distance between them (1.5×10^{11} m).

9.2 A: $|\vec{F_g}| = G\dfrac{m_1 m_2}{r^2} = \left(6.67 \times 10^{-11} \; {}^{N \cdot m^2}/_{kg^2}\right) \dfrac{(6 \times 10^{24} \; kg)(2 \times 10^{30} \; kg)}{(1.5 \times 10^{11} \; m)^2} = 3.5 \times 10^{22} \; N$

As you can see, the force of gravity is a relatively weak force, and you would expect a relatively weak force between relatively small objects. It takes tremendous masses and relatively small distances in order to develop significant gravitational forces. Let's take a look at another problem to explore the relationship between gravitational force, mass, and distance.

9.3 Q: As a meteor moves from a distance of 16 Earth radii to a distance of 2 Earth radii from the center of Earth, the magnitude of the gravitational force between the meteor and Earth becomes
A) one-eighth as great
B) 8 times as great
C) 64 times as great
D) 4 times as great

9.3 A: (C) 64 times as great. The gravitational force is given by Newton's Law of Universal Gravitation. If the radius is one-eighth its initial value, and radius is squared in the denominator, the radius squared becomes one-sixty fourth its initial value. Because radius squared is in the denominator, the gravitational force must increase by 64X.

9.4 Q: The diagram below represents two satellites of equal mass, A and B, in circular orbits around a planet.

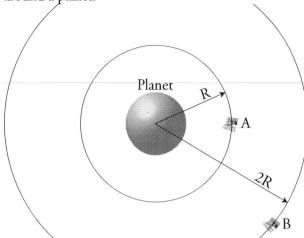

Compared to the magnitude of the gravitational force of attraction between satellite A and the planet, the magnitude of the gravitational force of attraction between satellite B and the planet is

A) half as great

B) twice as great

C) one-fourth as great

D) four times as great

9.4 A: (C) one-fourth as great. As distance between the objects' centers doubles, the gravitational force is quartered consistent with the inverse square law relationship.

Gravitational Fields

Gravity is a non-contact, or field, force. Its effects are observed without the two objects coming into contact with each other. Exactly how this happens is a mystery to this day, but scientists have come up with a mental construct known as a vector field to help understand how gravity and other field forces work.

A vector field describes the value of some physical vector quantity at a specific point in time in space. Envision an object with a gravitational field, such as the planet Earth. The closer other masses are to Earth, the more gravitational force they will experience. You can characterize this by calculating the amount of force the Earth will exert per unit mass at various distances from the Earth. For example, right now, at the location where you are sitting on the surface of the Earth, the gravitational field strength (**g**) is approximately 9.8 newtons of force toward the center of the Earth for every kilogram of gravitational mass, or 9.8 N/kg. If your gravitational mass is 60kg, then the gravitational force you are experiencing is (60kg)×(9.8 N/kg) or 588 newtons down. More commonly, you refer to this as your weight.

Obviously, the closer an object is to the Earth, the larger a gravitational force it will experience, and the farther it is from the Earth, the smaller a gravitational force it will experience.

Chapter 9: Gravity

Attempting to visualize this, picture the strength of the gravitational force on a test object represented by a vector arrow at the position of the object. The denser the force vectors are, the stronger the force, the stronger the "gravitational field." As these field lines become less dense, the gravitational field gets weaker.

If more than one source of a vector field exists, the net field value at a particular point in time and space can be determined by vector addition of vector fields due to each individual source. Visualize a universe in which only two suns exist. Determining the net gravitational field anywhere in the universe would be as simple as determining the gravitational field due to the first sun at the point in space, determining the gravitational field due to the second sun at the point in space, and then adding up the two gravitational field vectors.

A gravitational field (**g**) acting on an object of mass m produces a gravitational force of attraction mg in the direction of the gravitational field vector. In mathematical terms:

$$\vec{F}_g = m\vec{g} \rightarrow \vec{g} = \frac{\vec{F}_g}{m}$$

Combining this equation with Newton's Law of Universal Gravitation, you can derive a general relationship for the gravitational field strength outside a spherically symmetric object.

$$|\vec{g}| = \frac{|\vec{F}_g|}{m} = \frac{G\frac{m_1 m_2}{r^2}}{m} = G\frac{m}{r^2}$$

In the region near Earth's surface, the gravitational field strength **g** is relatively constant. This has allowed us to approximate the gravitational force, or weight, of objects near the surface of the earth as *mg* up to this point. This approximation breaks down if the gravitational field strength is no longer constant, such as when an object is positioned far from the surface of the Earth.

But wait, you might say — I thought **g** was the acceleration due to gravity on the surface of the Earth! And you would be right. Not only is **g** the gravitational field strength, it's also the acceleration due to gravity. The units even work out. The units of gravitational field strength, N/kg, are equivalent to the units for acceleration, m/s²!

Still skeptical? Try to calculate the gravitational field strength on the surface of the Earth using the knowledge that the mass of the Earth is approximately 5.98×10^{24} kg and the distance from the surface to the center of mass of the Earth (which varies slightly since the Earth isn't a perfect sphere) is approximately 6378 km.

$$\vec{g} = G\frac{m}{r^2} = \left(6.67 \times 10^{-11} \, {}^{N \cdot m^2}\!/_{kg^2}\right)\frac{(5.98 \times 10^{24} \, kg)}{(6378000 \, m)^2} = 9.8 \, {}^{m}\!/_{s^2}$$

As expected, the gravitational field strength on the surface of the Earth is the acceleration due to gravity.

9.5 Q: Suppose a 100-kg astronaut feels a gravitational force of 700N when placed in the gravitational field of a planet.
A) What is the magnitude of the gravitational field strength at the location of the astronaut?
B) What is the mass of the planet if the astronaut is 2×10^6 m from its center?

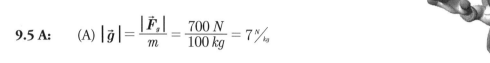

9.5 A: (A) $|\vec{g}| = \frac{|\vec{F}_g|}{m} = \frac{700 \, N}{100 \, kg} = 7 \, {}^{N}\!/_{kg}$

(B) $|F_g| = G\frac{m_1 m_2}{r^2} \rightarrow m_2 = \frac{F_g r^2}{Gm_1} = \frac{(700 \, N)(2 \times 10^6 \, m)}{(6.67 \times 10^{-11} \, {}^{N \cdot m^2}\!/_{kg^2})(100kg)} = 4.2 \times 10^{23} \, kg$

9.6 Q: What is the acceleration due to gravity at a location where a 15-kilogram mass weighs 45 newtons?
(A) 675 m/s²
(B) 9.81 m/s²
(C) 3.00 m/s²
(D) 0.333 m/s²

9.6 A: (C) $g = \frac{F_g}{m} = \frac{45N}{15kg} = 3 \, {}^{N}\!/_{kg} = 3 \, {}^{m}\!/_{s^2}$

9.7 Q: A baseball is dropped from a height of 5,000 km above Earth's surface. Determine the initial acceleration of the baseball.

9.7 A: $|\vec{g}| = \frac{Gm}{r^2} = \frac{(6.67 \times 10^{-11} \, {}^{N \cdot m^2}\!/_{kg^2})(6 \times 10^{24} kg)}{(5,000,000 \, m + 6.37 \times 10^6 \, m)^2} = 3.1 \, {}^{m}\!/_{s^2}$

9.8 Q: A planet with the same mass volume mass density as Earth but half the radius would have what acceleration due to gravity at its surface?

9.8 A: The new planet would have one-eighth the volume (and therefore one-eighth the mass) of Earth since volume is proportional to r³, resulting in one-eighth the acceleration. Additionally, the new planet has half the radius of Earth, which will give it four times the acceleration due to gravity due to the inverse square law relationship. Putting these effects together, you get one half the acceleration due to gravity on Earth, or 4.9 m/s².

Chapter 9: Gravity

Gravitational Fields In Shells and Spheres

We've treated gravitational fields outside uniform solid spheres as if all the mass of the sphere were concentrated at a singular point in the center of the sphere. This works very well as long as you're looking for the gravitational field outside the sphere. But what would change if you wanted to know the gravitational field strength somewhere inside the sphere? What would happen if instead of a uniform solid sphere, the sphere was hollow?

Inside a hollow sphere, the gravitational field is zero. Outside a hollow sphere, you can treat the sphere as if its entire mass is concentrated at the center, and then calculate the gravitational field.

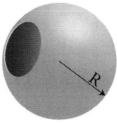
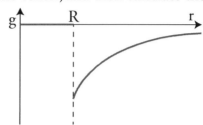

9.9 Q: A hollow spherical shell has a mass of 2500 kg and a radius of 5 m. Determine the gravitational field strength at each of the following distances from the center of the shell.
A) 2 m
B) 4 m
C) 6 m

9.9 A: A) 0
B) 0

$$C) \quad g(6m) = \frac{Gm}{r^2} = \frac{(6.67 \times 10^{-11})(2500)}{6^2} = 4.63 \times 10^{-9}\,{}^{N}\!/_{kg}$$

Outside a solid sphere, treat the sphere as if all the mass is at the center of the sphere. Inside the sphere, treat the sphere as if the mass inside the radius is all at the center. Only the mass inside the "radius of interest" counts in your gravitational calculations. For a uniform volume density, inside the sphere the gravitational field strength will be proportional to the distance from the sphere's center, and outside the sphere the gravitational field strength will again be inversely proportional to the square of the distance from the sphere's center.

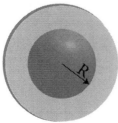
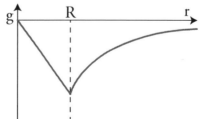

9.10 Q: A uniform solid sphere has a mass of 2500 kg and a radius of 5 m. Determine the gravitational field strength at each of the following distances from the center of the shell.
A) 4 m
B) 6 m

9.10 A: A) First determine the volume mass density of the sphere:

$$\rho = \frac{M}{V} = \frac{M}{\frac{4}{3}\pi r^3} = \frac{3M}{4\pi r^3} = \frac{3(2500)}{4\pi (5^3)} = 4.77 \, \text{kg}/\text{m}^3$$

Next, find the mass inside the radius of interest, 4m.

$$M_{enclosed} = \rho V = 4.77\left(\frac{4}{3}\pi (4)^3\right) = 1280 \, kg$$

Finally, solve for the gravitational field strength using the mass inside the radius of 4 m.

$$g(4m) = \frac{Gm}{r^2} = \frac{(6.67 \times 10^{-11})(1280)}{4^2} = 5.34 \times 10^{-9} \, N/kg$$

B) $$g(6m) = \frac{Gm}{r^2} = \frac{(6.67 \times 10^{-11})(2500)}{6^2} = 4.63 \times 10^{-9} \, N/kg$$

Gravitational Potential Energy

Previously you learned that gravitational potential energy is the energy an object possesses because of its position in a gravitational field, and performed some basic calculations involving constant gravitational fields. You can generalize and expand this relationship by tying in Newton's Law of Universal Gravitation to find the gravitational potential energy existing due to the interaction of any two objects.

In exploring gravitational potential energy, we have previously used mgh for our calculations, assuming the gravitational field strength near the surface of the Earth is constant. As you move away from Earth's surface, however, this is no longer the case, and a more general expression for gravitational potential energy is required. We'll set a reference value of zero gravitational potential energy an infinite distance away from all other objects. As gravity is a conservative force, we can find the gravitational potential energy from the definition of potential energy.

$$dU = -\vec{F}\cdot d\vec{l} \xrightarrow{d\vec{l}=d\vec{r}} dU = -\vec{F}\cdot d\vec{r} \rightarrow \int dU = \int_{r=\infty}^{r} -\vec{F}\cdot d\vec{r} \rightarrow U = -\int_{r=\infty}^{r} F dr \xrightarrow{F=-G\frac{m_1 m_2}{r^2}}$$

$$U = -\int_{r=\infty}^{r} -G\frac{m_1 m_2}{r^2} dr \rightarrow U = Gm_1 m_2 \int_{r=\infty}^{r} r^{-2} dr = Gm_1 m_2 \left(\frac{-1}{r}\right)\Big|_{\infty}^{r} = Gm_1 m_2 \left(\frac{-1}{r} - \frac{-1}{\infty}\right) \rightarrow$$

$$U = -\frac{Gm_1 m_2}{r}$$

Note the negative value for the gravitational potential energy. As gravitational potential energy is a relative measure, the fixed reference point of zero gravitational potential energy is set when objects are an infinite distance apart from each other. As the objects get closer together, the magnitude of their gravitational potential energy increases, but is defined as a negative, indicating that the objects are "caught" in each other's gravitational field. Therefore, absolute gravitational potential energies are negative by convention.

9.11 Q: A binary star is a celestial phenomenon in which two stars orbit around their mutual center of mass. If the mass of the primary star is 4M, and the mass of the secondary star is M, what is the gravitational potential energy of the binary star system if the stars are separated by distance r?

9.11 A: $U_g = -\dfrac{Gm_1 m_2}{r} = -\dfrac{G(4M)(M)}{r} = -\dfrac{4GM^2}{r}$

9.12 Q: A rocket is launched vertically from the surface of the Earth with an initial velocity of 10 km/s. What maximum height does the rocket reach, neglecting air resistance? Note that the mass of the Earth is 6×10^{24} kg and the radius of the Earth is 6.37×10^6 m. You may not assume that the acceleration due to gravity is a constant.

9.12 A: Using a conservation of energy approach, the initial total mechanical energy at the time of the rocket's launch must equal the rocket's total mechanical energy when it reaches its highest point. Therefore you can write $K_i + U_i = K_f + U_f$.

The problem then becomes an exercise in algebra, solving for R at the highest point of the rocket's flight and recognizing that the rocket has a kinetic energy of 0 at its highest point. Call the mass of the Earth m_1, the mass of the rocket m_2, the radius of the Earth R_E, and the radius of the rocket R.

$$\tfrac{1}{2}m_2 v^2 + \frac{-Gm_1 m_2}{R_E} = 0 + \frac{-Gm_1 m_2}{R} \rightarrow \frac{1}{R} = \frac{-m_2 v^2}{2Gm_1 m_2} + \frac{Gm_1 m_2}{Gm_1 m_2 R_E} \rightarrow$$

$$\frac{1}{R} = \frac{-v^2}{2Gm_1} + \frac{1}{R_E} \rightarrow \frac{1}{R} = \frac{-(10,000)^2}{2(6.67 \times 10^{-11})(6 \times 10^{24})} + \frac{1}{6.37 \times 10^6} \rightarrow R = 3.12 \times 10^7 \ m$$

Having now found R, the maximum distance the rocket reaches from the center of the Earth, you can solve for h, the height of the rocket above Earth's surface.

$h = R - R_E = 3.12 \times 10^7 \ m - 6.37 \times 10^6 \ m = 2.48 \times 10^7 \ m$

Orbits

How do celestial bodies orbit each other? The moon orbits the Earth. The Earth orbits the sun. Earth's solar system is in orbit in the Milky Way galaxy... but how does it all work?

To explain orbits, Sir Isaac Newton developed a "thought experiment" in which he imagined a cannon placed on top of a very tall mountain, so tall, in fact, that the peak of the mountain was above the atmosphere (this is important because it allows us to neglect air resistance). If the cannon then launched a projectile horizontally, the projectile would follow a parabolic path to the surface of the Earth.

If the projectile was launched with a higher speed, however, it would travel farther across the surface of the Earth before reaching the ground. If its speed could be increased high enough, the projectile would fall at the same rate the Earth's surface curves away. The projectile would continue falling forever as it circled the Earth! This circular motion describes an orbit.

Put another way, the astronauts in the space shuttle aren't weightless. Far from it, actually; the Earth's gravity is still acting on them and pulling them toward the center of the Earth with a substantial force. You can even calculate that force and the acceleration due to gravity while they're in orbit.

9.13 Q: If the space shuttle orbits the Earth at an altitude of 380 km, what is the gravitational field strength due to the Earth at that altitude? Mass of the Earth is 5.98×10^{24} kg, and the radius of the Earth is approximately 6.37×10^6 m.

9.13 A: $F_g = mg = \dfrac{Gm_1 m_2}{r^2} \rightarrow g = \dfrac{Gm_{Earth}}{r^2} = \dfrac{(6.67 \times 10^{-11})(5.98 \times 10^{24})}{(6.37 \times 10^6 + 380 \times 10^3)} = 8.75 \, {}^N\!/\!_{kg} = 8.75 \, {}^m\!/\!_{s^2}$

This means that the acceleration due to gravity at the altitude the astronauts are orbiting the earth is only 11% less than on the surface of the Earth! In actuality, the space shuttle is falling, but it's moving so fast horizontally that by the time it falls, the Earth has curved away underneath it so that the shuttle remains at the same distance from the center of the Earth. It is in orbit! Of course, this takes tremendous speeds. To maintain an orbit of 380 km, the space shuttle travels approximately 7680 m/s, more than 23 times the speed of sound at sea level!

An object orbiting a celestial body such as a planet in a circular orbit must be held in its circular path by a centripetal force. The centripetal force causing this circular motion is, of course, gravity. You can use this knowledge to determine the speed of any satellite in a circular orbit as a function of the mass of the object it is orbiting and the radius of its orbit.

You can apply your knowledge of gravity and circular motion to find the velocity of any circular orbit. Consider a mass m_2 in a circular orbit with tangential velocity v about mass m_1 as shown below.

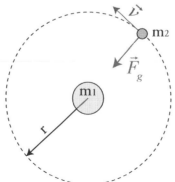

Starting with Newton's 2nd Law and recognizing the centripetal force is the force of gravity, solve for the velocity.

$$F_{net_c} = ma_c = F_g \xrightarrow[\substack{a_c = \frac{v^2}{r}}]{\substack{F_g = G\frac{m_1 m_2}{r^2}}} m_2 \frac{v^2}{r} = G\frac{m_1 m_2}{r^2} \rightarrow v = \sqrt{\frac{Gm_1}{r}}$$

Geosynchronous orbit occurs when a satellite maintains an orbit which has a period of one day, so that at the same time each day, the satellite is above the same point on the Earth. **Geostationary orbits** are orbits directly above the equator, which allow satellites to remain above the same point on the Earth (the satellite's rotational velocity matches the Earth's rotational velocity). These are very useful for communications satellites. To "speed up" or "slow down" an orbiting satellite relative to the surface of the Earth, you just change the satellite's altitude.

9.14 Q: Given the mass of the Earth is approximately $6{\times}10^{24}$ kg, and the radius of the Earth is approximately $6.37{\times}10^{6}$ m, determine the speed and the height above the surface of the Earth for a satellite in a geostationary orbit.

9.14 A: For an object in geostationary orbit, the period of its orbit must be 24 hours, or 86,400 seconds. Using this, you can develop an equation for the speed of the satellite as a function of R: $v = \dfrac{2\pi R}{T}$.

Next, utilize the relationship you developed from the previous problem between the velocity of a satellite and its orbital radius to solve for the velocity of the satellite.

$$v = \sqrt{\frac{Gm_1}{R}} \xrightarrow[R=\frac{vT}{2\pi}]{v=\frac{2\pi R}{T}} v = \sqrt{\frac{2\pi Gm_1}{vT}} \rightarrow v = \sqrt[3]{\frac{2\pi Gm_1}{T}} \rightarrow$$

$$v = \sqrt[3]{\frac{2\pi\left(6.67\times10^{-11}\right)\left(6\times10^{24}\right)m_1}{86,400}} = 3080 \text{ m}/_s$$

Then, solve for the radius of the satellite.

$$R = \frac{vT}{2\pi} = \frac{(3080)(86400)}{2\pi} = 4.2\times10^7 \text{ m}$$

Finally, solve for the height above the surface of the Earth by subtracting the radius of the Earth from the radius of the satellite's orbit.

$$h = R - R_E = 4.2\times10^7 - 6.37\times10^6 = 3.6\times10^7 \text{ m}$$

Escape velocity is the velocity required for an object to completely escape the influence of gravity, which theoretically occurs at an infinite distance from all other masses. You can find the escape velocity of an object by recognizing that the gravitational potential energy of the system is zero when an object has completely escaped gravity's influence, and it has no leftover kinetic energy (its speed is zero).

9.15 Q: Determine the escape velocity for an object of mass m_2 initially held at a distance R from the center of mass of a large celestial body of mass m_1.

9.15 A: Begin by recognizing that the total energy of the system is zero when there is no gravitational potential energy (no gravitational influence) and the object's kinetic energy is zero (speed is zero). Use this relationship to solve for the initial velocity required to reach this state.

$$E_{total} = K + U_g = 0 \rightarrow \frac{1}{2}m_2v^2 - G\frac{m_1 m_2}{r} = 0 \rightarrow \frac{1}{2}m_2v^2 = G\frac{m_1 m_2}{r} \rightarrow v^2 = \frac{2Gm_1 m_2}{m_2 r}$$

$$\rightarrow v_{escape} = \sqrt{\frac{2Gm_1}{r}}$$

9.16 Q: Determine the total mechanical energy for a satellite of mass m_2 in orbit around a much larger object of mass m_1 in terms of the two masses and the distance between their centers of mass.

9.16 A: Begin by recognizing that the total mechanical energy is the sum of the kinetic and potential energies. Next, substitute in expressions for the kinetic and gravitational potential energies. Finally, utilize the relationship for the velocity of a satellite in orbit to remove the equation's dependence on v and solve for the total energy.

$$E = K + U = \frac{1}{2}m_2 v^2 - G\frac{m_1 m_2}{r} \xrightarrow[v^2 = \frac{Gm_1}{r}]{v = \sqrt{\frac{Gm_1}{r}}} E = \frac{1}{2}m_2\frac{Gm_1}{r} - G\frac{m_1 m_2}{r} \rightarrow E = -G\frac{m_1 m_2}{2r}$$

Kepler's Laws of Planetary Motion

In the early 1600s, most of the scientific world believed that the planets should have circular orbits, and many believed that the Earth was the center of the solar system. Using data collected by Tycho Brahe, German astronomer Johannes Kepler developed three laws governing the motion of planetary bodies, which described their orbits as ellipses with the sun at one of the focal points (even though the orbits of many planets are nearly circular). These laws are known as **Kepler's Laws of Planetary Motion.**

Kepler's First Law of Planetary Motion states that the orbits of planetary bodies are ellipses with the sun at one of the two foci of the ellipse.

Kepler's Second Law of Planetary Motion states that if you were to draw a line from the sun to the orbiting body, the body would sweep out equal areas along the ellipse in equal amounts of time. This is easier to observe graphically. In the diagram below, if the orbiting body moves from point 1 to point 2 in the same amount of time as it moves from point 3 to point 4, then Area 1 and Area 2 must have the same shaded area.

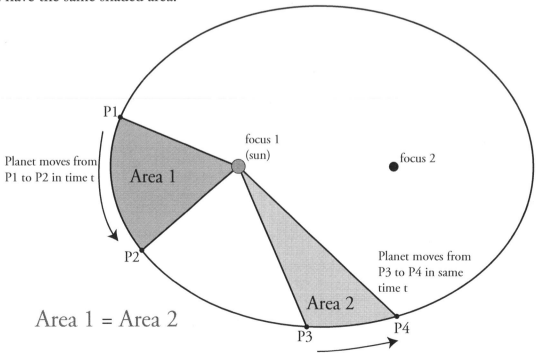

9.17 Q: Given the elliptical planetary orbit shown in the previous diagram, identify the interval during which the planet travels with the highest speed.
A) Interval P1 to P2
B) Interval P3 to P4
C) They are the same.

9.17 A: (A) Interval P1 to P2. Because Area 1 is equal to Area 2, we know that the time interval from P1 to P2 must be equal to the time interval from P3 to P4 by Kepler's 2nd Law of Planetary Motion. Since the planet travels a greater distance from P1 to P2, it must have the higher speed during this portion of its journey.

Note that Kepler's 2nd Law of Planetary Motion is really an application of the law of conservation of angular momentum. If you analyze the system from the perspective of reference point P as shown in the diagram below, and there are no external forces or torques (the only force is gravity through point P), then angular momentum must be conserved.

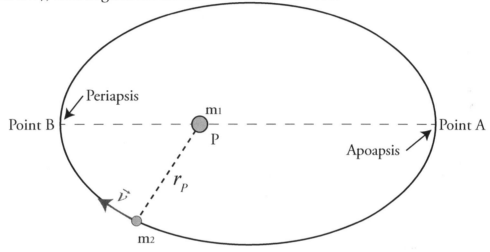

You can then write a conservation of angular momentum equation to obtain a relationship for velocity and radius for any point in orbit:

$$|\vec{L}_P| = mvr\sin\theta \rightarrow m_2 v_A r_A \sin\theta_A = m_2 v_B r_B \sin\theta_B$$

Point A on the diagram, in which the orbiting body is furthest from the object it is orbiting about, is known as **apoapsis**. Point B, in which the orbiting body is closest to the object it is orbiting about, is known as **periapsis**. At points A and B, $\theta=90°$, so the analysis for these specific points simplifies to:

$$v_A r_A = v_B r_B$$

Consider mass m_1 as the sun and mass m_2 as an orbiting planet. The planet is farther from the sun at point A (apogee), and therefore moves slower than when it is at point B (perigee).

Author's Note: Astronomers use more specific terms for apoapsis and periapsis coinciding with the specific body the object is orbiting about. Objects orbiting the Earth, for example, have points in their orbits known as **apogee** and **perigee**. Objects orbiting the Sun have points known as **aphelion** and **perihelion**. A simple tool for remembering which term applies to which position involves noting the first letter of the term. Apoapsis, apogee, and aphelion are all points that are furthest **away** from one of the foci of the orbital ellipse.

Kepler's 3rd Law of Planetary Motion, described several years after the first two laws were published, states that the ratio of the squares of the periods of two planets is equal to the ratio of the cubes of their orbital radii. If this sounds confusing, don't worry! Once again it's not as bad as it looks, and is something that is quite easy to derive. In deriving Kepler's 3rd Law, we'll have to make an assumption that planets move in circular orbits. This is a reasonable approximation for the planets closest to our sun.

The period (T) is the time it takes for a satellite to make one complete revolution, or travel one circumference, of its circular path. You can solve for the period using the velocity for a satellite in orbit that you found earlier in this chapter.

$$2\pi R = vT \rightarrow T = \frac{2\pi R}{v} \xrightarrow{v=\sqrt{\frac{Gm_1}{R}}} T = \frac{2\pi R\sqrt{R}}{\sqrt{Gm_1}}$$

Next, you can clean up the expression by squaring both sides and dividing by R^3.

$$T = \frac{2\pi R\sqrt{R}}{\sqrt{Gm_1}} \rightarrow T^2 = \frac{4\pi^2 R^3}{Gm_1} \rightarrow \frac{T^2}{R^3} = \frac{4\pi^2}{Gm_1}$$

This ratio, T^2/R^3, is a constant for any orbit! In our solar system, for example, T^2/R^3 is approximately equal to 2.97×10^{-19} s²/m³.

What this is really saying, then, is that planets that are closer to the sun (with a smaller orbital radius) have much shorter periods than planets that are farther from the sun. For example, the planet Mercury, closest to the sun, has an orbital period of 88 days. Neptune, which is 30 times farther from the sun than Earth, has an orbital period of 165 Earth years.

9.18 Q: A satellite orbits a planet in an elliptical path as shown. Specific positions of the satellite are noted on the diagram as A, B, C, and D.

Rank from highest to lowest the following characteristics of the satellite at each position.
I) Speed
II) Gravitational Potential Energy
III) Total Mechanical Energy

9.18 A: I) Speed: C, D, A, B (Kepler's 2nd / 3rd Laws)
II) U_g: B, A, D, C (law of conservation of energy)
III) E_T: same for all (law of conservation of energy)

9.19 Q: Given the orbital radius of Mercury is roughly $5.8{\times}10^{10}$ m, estimate the period of Mercury's orbit in terms of Earth years.

9.19 A: First, utilize Kepler's 3rd Law to find the ratio of T^2/R^3.
$$\frac{T^2}{R^3} = \frac{4\pi^2}{Gm_1} = \frac{4\pi^2}{(6.67 \times 10^{-11})(2 \times 10^{30})} = 2.9 \times 10^{-19} \, {}^{s^2}\!/_{m^3}$$

Next, solve for the period of Mercury's orbit in seconds.
$$T^2 = R^3(2.96 \times 10^{-19}) = (5.8 \times 10^{10})^3(2.96 \times 10^{-19}) = 5.77 \times 10^{13} \to T = 7.6 \times 10^6 \, s$$

Finally, convert Mercury's period in seconds into Earth years.
$$7.6 \times 10^6 \, s \times \frac{1 \, yr}{3.16 \times 10^7 \, s} = 0.24 \, yr$$

9.20 Q: The shape of Mars' orbit around the sun is most accurately described as a:
A) circle
B) ellipse
C) parabola
D) hyperbola

9.20 A: (B) ellipse. The orbits of planets are ellipses with the sun as one of the foci of the ellipse. Note that even though the orbits are best described as ellipses, many of the planetary orbits are "nearly circular."

We've applied Kepler's 3rd Law of Planetary Motion to circular orbits, but in actuality, the planetary orbits are ellipses, even if they are close to circular. This law can be generalized for elliptical orbits, where the ratio of the squares of the periods of two planets is equal to the ratio of the cubes of their semi-major axes: $\frac{T^2}{a^3} = \frac{4\pi^2}{Gm_1}$. This ratio of the squares of the periods to the cubes of the semi-major axes is referred to as Kepler's Constant.

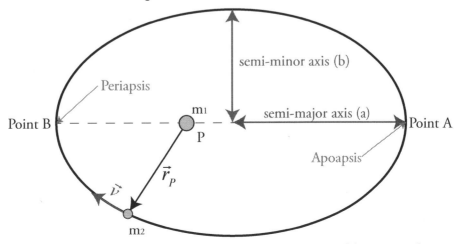

The total mechanical energy for an elliptical orbit is then given by $E = -G\dfrac{m_1 m_2}{2a}$.

9.21 Q: A satellite of mass m_s orbits a planet of mass m_p at an altitude equal to twice the radius (R) of the planet. What is the satellite's speed assuming a perfectly circular orbit?

A) $v = \sqrt{\dfrac{Gm_p}{R}}$

B) $v = \sqrt{\dfrac{Gm_s}{R}}$

C) $v = \sqrt{\dfrac{Gm_s}{2R}}$

D) $v = \sqrt{\dfrac{Gm_p}{3R}}$

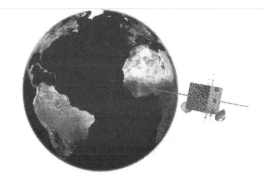

9.21 A: (D). First recognize that the radius of the satellite's orbit is 3R, the radius of the planet plus the altitude of the satellite above the surface of the planet. Then, a force analysis recognizing the gravitational force of attraction provides a centripetal force yields:

$$F_{net_c} = F_g = ma_c = \frac{mv^2}{r} \rightarrow \frac{m_s v^2}{3R} = \frac{Gm_s m_p}{(3R)^2} \rightarrow v^2 = \frac{Gm_p}{3R} \rightarrow v = \sqrt{\frac{Gm_p}{3R}}$$

9.22 Q: A spaceship in a circular orbit 400 km above the surface of the Earth wishes to manipulate its orbit to reach a point P on the opposite side of the Earth which is 1000 km above the Earth's surface. If the spaceship is at the position shown in the diagram and currently moving in a clockwise direction, in which direction should the ship accelerate in order to reach point P?
A) toward the top of the page
B) toward the right of the page
C) toward the bottom of the page
D) toward the left of the page

9.22 A: (B) toward the right of the page. To increase the radius of its orbit, the ship must attain a higher velocity, which requires an acceleration in the direction of its current velocity, or to the right of the page as depicted in this diagram. This will shift the orbit from a circular orbit to an elliptical orbit, and allow the ship to reach point P. (Note that a second acceleration will be required upon reaching point P if the ship wishes to maintain a circular orbit 1000 km above the Earth's surface.)

9.23 Q: A rock is thrown horizontally from the top of a 100-meter-high vertical cliff on Planet Unicorn with a speed of 20 m/s. If the mass of Planet Unicorn is 1025 kg and the top of the cliff is approximately 4000 kilometers from the center of the planet, how far from the base of the cliff does the rock land?

9.23 A: The acceleration due to gravity is the gravitational field strength, which can be determined from Newton's Law of Universal Gravitation. The horizontal distance traveled by the rock is the time it takes for the rock to strike the ground (a kinematics problem) multiplied by the horizontal velocity of the rock.

$$\Delta x = v_x t \xrightarrow[t=\sqrt{\frac{2h}{g}}]{v_x=20} \Delta x = 20\sqrt{\frac{2h}{g}} \xrightarrow{g=\frac{Gm}{r^2}} \Delta x = 20\sqrt{\frac{2hr^2}{Gm}} \rightarrow$$

$$\Delta x = 20\sqrt{\frac{2(100)(4\times10^6)^2}{(6.67\times10^{-11})(10^{25})}} = 43.8\ m$$

9.24 Q: Which of the following changes would increase the magnitude of the gravitational field intensity an object feels when near a planet? (Select two answers.)
A) increase the mass of the object
B) increase the mass of the planet
C) decrease the spin rate of the planet
D) decrease the separation distance between object and planet

9.24 A: (B) and (D). By definition, field intensity, $g = \dfrac{F_g}{m_{object}}$, expands to $g = \dfrac{Gm_{planet}}{r^2}$ when

Newton's Law of Universal Gravitation is applied, leading to choices (B) and (D).

9.25 Q: Marty is an astronaut who is preparing to go on a mission in orbit around the Earth. For health reasons, his mass needs to be determined before take-off and while he is in orbit. The morning of the launch, Marty sits on one pan of a two-pan scale and 94 kg of mass is needed to balance him.

A) State and explain whether the two-pan scale registered Marty's gravitational mass or inertial mass.

B) After a few days in orbit, Marty is again able to determine his mass. Explain why the two-pan scale used before launch cannot be used to measure his mass while in orbit.

C) To determine Marty's mass in orbit, he is to sit in a chair of negligible mass that is attached to a wall by a spring that has a force constant, k. Consequently, the chair freely vibrates back and forth with a period, T, when displaced sideways a distance x. Explain how the spring-mounted chair can be used to determine Marty's mass, m. Give relevant measurements and equation(s).

D) If Marty has lost mass while in orbit, what specific change would occur when he sits in the chair and starts it oscillating?

E) Explain why this spring-mounted chair measures Marty's inertial mass.

9.25 A: A) Gravitational mass. The pans of the scale balance under the influence of gravity, not any other force.

B) In orbit the effects of gravity are not felt because everything is in free-fall. Consequently, the scales will not balance when objects are placed on them.

C) Marty's mass can be determined by measuring the period of vibration of the oscillating chair (displacement is irrelevant) and using the equation $T = 2\pi \sqrt{\frac{m}{k}}$.

D) The period of oscillation would decrease (no change in displacement).

E) Inertial mass affects an object's response to a non-gravitational force as described by Newton's 2nd Law of Motion. In this situation, the force is due to the spring and the response is the period of oscillation.

9.26 Q: A space probe of negligible mass is sent on a mission to map out the gravitational field intensity in the vicinity of a satellite of planet X. Some of the data collected is shown in the chart below.

distance to satellite (×10⁶ m)	1/R² (1/m²)	field intensity (N/kg)
2.0		13.3
2.5		8.4
3.0		5.9
3.5		4.5
4.0		3.3
6.0		1.5

A) Plot the gravitational field intensity (g) vs. the distance (R) to the satellite.

B) Draw the appropriate best fit line or curve.

C) Using the best fit, what distance corresponds to a field intensity of 2.1 N/kg?

D) In order to determine the mass of the satellite, a plot of field intensity vs. 1/R² can be utilized. Fill in the appropriate values for 1/R² in the chart above.

E) Plot gravitational field intensity, g, on the y-axis vs. 1/R² on the x-axis.

F) Use the plot and best fit to determine the mass of the planet.

9.26 A: A) Draw plot.

B) Draw best-fit curve.

C) 5.0×10^6 m

D)

1/R² (1/m²)	field intensity (N/kg)
2.5×10^{-13}	13.3
1.6×10^{-13}	8.4
1.1×10^{-13}	5.9
8.2×10^{-14}	4.5
6.3×10^{-14}	3.3
2.8×10^{-14}	1.5

E) Draw plot.

F) slope=$gR^2 = 5 \times 10^{13}$ m³/s². If $g = Gm_p/R^2$ then $m_p = (gR^2)/G = $ slope/G $ = 7.5 \times 10^{23}$ kg

Index

Made in the USA
Lexington, KY
16 October 2018